I0042449

CAPITAL AND INEQUALITY IN RURAL PAPUA NEW GUINEA

CAPITAL AND INEQUALITY IN RURAL PAPUA NEW GUINEA

Edited by Bettina Beer
and Tobias Schwoerer

Australian
National
University

ANU PRESS

ASIA-PACIFIC ENVIRONMENT MONOGRAPH 16

Australian
National
University

ANU PRESS

Published by ANU Press
The Australian National University
Canberra ACT 2600, Australia
Email: anupress@anu.edu.au

Available to download for free at press.anu.edu.au

ISBN (print): 9781760465186
ISBN (online): 9781760465193

WorldCat (print): 1331091831
WorldCat (online): 1331091830

DOI: 10.22459/CIRPNG.2022

This title is published under a Creative Commons Attribution-NonCommercial-NoDerivatives 4.0 International (CC BY-NC-ND 4.0) licence.

The full licence terms are available at
creativecommons.org/licenses/by-nc-nd/4.0/legalcode

Cover design and layout by ANU Press. Cover photograph: Wampar houses (B. Beer and T. Schwoerer).

This book is published under the aegis of the Asia-Pacific Environment Monographs Editorial Board of ANU Press.

This edition © 2022 ANU Press

Contents

Contributors

Glenn Banks is a professor and head of the School of People, Environment and Planning at Massey University: Te Kunenga ki Pūrehuroa, New Zealand.

Bettina Beer is a professor and head of the Department of Social and Cultural Anthropology at the University of Lucerne, Switzerland.

Willem Church is a research fellow at the Max Planck Institute for Evolutionary Anthropology in Leipzig, Germany.

Peter D. Dwyer is an honorary research fellow at the University of Melbourne.

Bruce Knauft is the Samuel C. Dobbs Professor of Anthropology at Emory University.

Monica Minnegal is an associate professor of anthropology at the University of Melbourne.

Tobias Schwoerer is a senior lecturer in the Department of Social and Cultural Anthropology at the University of Lucerne, Switzerland.

List of Figures and Tables

1

Capital and Inequality in Rural Papua New Guinea

Bettina Beer and Tobias Schwoerer[1]

Introduction

As international capital inserts itself across the Pacific, its benefits and burdens tend to be unequally distributed among governments, corporations and different groupings of local people. The emergence of inequality is clear enough in itself—the ongoing conflict and controversy surrounding the distribution of gains from extractive and other capital-intensive projects and their negative social and environmental impacts speak to this. However, there is ambiguity in how capital-intensive projects, coupled with the social contexts and pre-existing inequalities in which they operate, shape the form, magnitude and persistence of these inequalities.

In this edited volume, we will present accounts of how capital-intensive projects in the mining, oil and gas, and agro-industry sectors unfold to generate specific inequalities across diverse settings in rural and semi-rural Papua New Guinea (PNG). We focus on the beginnings of such

1   We thank the authors of the chapters in this collection, as well as the many colleagues who contributed to this volume. We would like to give special thanks to Doris Bacalzo and Don Gardner for helpful discussions before, during and after fieldwork, Bruce Knauft and Glenn Banks for workshops in Lucerne, Colin Filer for a workshop on Zoom and his encouragement during the publishing process, Robin Hide for updates on Markham-related publications we would otherwise have overlooked, and Lea Helfenstein for the final formatting and correcting of the manuscripts.

projects, as imaginations and socio-political transformations in their anticipation often have major repercussions on the way they impact the people involved and set precedents for the distribution of gains and losses. The 'beginning' of a mine can stretch over decades of prospection and sales of the mine from one multinational corporation to the next before extraction even begins.

We thus follow the processes that reinforce existing inequalities, and create novel inequalities, through the presence of capital-intensive projects from their inception. We are also interested in how those inequalities become reproduced over generations. We intend to show that the complexities generated by each project and their interaction, in a regional context, pose challenges to interpretation that can only be handled through intensive, ongoing, long-term and longitudinal ethnographic investigations.

Studying Social Inequality

The study of social inequality at national and international levels focuses on capital, wealth and other economic factors relevant to 'life chances' in a globalised world. In such a world, these factors are also relevant to people's experience of stratification within a local setting (in conjunction with gender, ethnicity, age and other categorical distinctions). However, the relational characteristics of social fields and the positions of people within them (the various forms of 'cultural capital') are relevant to life chances, especially in non-market or only partially marketised economies. The significance of inequalities in social processes and human well-being have been contested since before the social sciences came into being, and they remain so (Payne 2017). Nevertheless, the extent of wealth inequalities and inequalities in mortality and longevity within and between nation states are well enough understood for these to become objects of policy for international institutions and nation states (Houweling et al. 2001; Soubbottina 2004; European Trade Union Institute 2012; Keeley 2015), as well as, more recently, for large transnational corporations (TNCs): for example, via concepts of 'corporate social responsibility' and 'sustainable development' (Rondinelli and Berry 2000; Jenkins and Yakovleva 2006; Gilberthorpe and Banks 2012).

The study of inequality and its modes, which used to be a staple of political anthropology (e.g. Berreman and Zaretsky 1981; Tilly 1998, 2001), suffered somewhat with the turn from 'grand narratives' and the 'hermeneutics of suspicion' that they entailed. Nevertheless, there has been some resistance to these post-structuralist perspectives and the views held to justify them. Furthermore, post-structuralist insights generate their own hermeneutics of suspicion in detecting the exclusions and discrimination tacitly incorporated into habitual modes of thought and analysis, which sometimes silence local voices or discount non-Western modes of agency. Thus, many older concerns relevant to issues of inequality are actually more prominent than formerly, not only because of the social ramifications of globalised capitalism, but also because of an increased sensitivity to the need to take account of the various modes of agency at work in the generation of inequality. The implication of broad and narrow social processes in the production and reproduction of inequalities therefore remains a focus for research. Theorising contemporary currents of globalisation and its significance for various kinds of inequality has, accordingly, remained a feature of the social sciences, including anthropology (e.g. Friedman 1994; Biersack and Greenberg 2006; Schuerkens 2010; Hann and Hart 2011; Friedman and Friedman 2013). In the ethnography of Melanesia, such concerns have also received careful treatment (e.g. Akin and Robbins 1999; Knauft 1999, 2002; LiPuma 2000; Robbins and Wardlow 2005; Wardlow 2006, 2020; West 2016; Bainton et al. 2021).

Anthropological theory and research have been integral to broader discussions about modernity and development, just as national and international interventions aimed at improving local conditions for communities 'in need' have always been of interest to anthropological theorists and ethnographic researchers. Today, both empirical research and anthropological high theory continue to address inequality and its relation to social conditions, be they pre-modern, modern or post-modern configurations (e.g. Scheper-Hughes 1993; Friedman 2000; Tsing 2005; Ferguson 2006; Li 2007; Smith et al. 2010). With the mistrust of older grand narratives there is a greater emphasis on social engagement, political commitment and amelioration of life conditions as a basis for research in the writings of sociocultural anthropologists, but older ideals of emancipatory critique grounded in systematic objectivist analysis continue to be represented in anthropology (e.g. Durrenberger 2012). Accordingly, concerns about inequality find expression in almost all parts of the contemporary anthropological scene, in explicit and implicit formulations. And with the increased theoretical stress on ethnography as the engine of anthropological thought, newer kinds of studies (of corporations and market mechanisms) have diversified the social domains and discursive fields relevant to issues of inequality.

The research results presented in this volume are focused upon the transformation of modes of inequality as a historically (relatively) non-market local economy is drawn into intimate and routine relations with the global circulation of capital implicated in large-scale extractive and large-scale plantation projects. All contributors seek to provide an ethnographic account of the processes that transform patterns of inequality among people indirectly and directly exposed to large-scale capital-intensive projects—such as a mine, a biomass or a liquid natural gas project—on a quotidian basis.

Ethnography of Small-Scale Life-Worlds

The research presented in this volume is a contribution to the anthropology of the encompassment of local, situated ways of life by institutions of globalised capital and the frictions thereby engendered (Tsing 2005). Broadly practice-theoretic, its focus is on the social micro-processes through which historically novel forms of inequality become entrenched under local manifestations of global capital imperatives. In that respect it is continuous with traditions of critical investigation and analysis that are as old as the social sciences, and evokes the 'Manchester School' strands in the anthropology of Africa that charted and questioned—if, often, only implicitly—the impact of European investments on local cultures (Burawoy 2000). However, today's globalised economy demands understandings of the relationship between inequality and 'community development' initiatives, for, in an officially decolonised world, state and international standards require assessments of large scale projects prior to and during their development (World Bank 2004). Anthropological and sociological investigations of large-scale developments reflect these interests (Kirsch 2006; Benson and Kirsch 2010; Shamir 2010; Bainton and Macintyre 2013; Banks et al. 2013; Gilberthorpe 2013a; Welker 2014).

Through this volume, we contribute to the understanding of the formation and transformation of local-level inequalities resulting from the anticipation of, and interaction and engagement with, large-scale globally financed projects. Filer (2007: 139–40) once likened projects of the 'Melanesian version of "heavy industry"' to a four-legged 'creature' or 'beast', composed of the mining industry, the oil and gas industry, the logging industry and the oil palm industry (or, in fact, any large-scale agro-industrial plantation industry). This volume explores three of the four industries: mining, oil and gas, and large-scale agro-industrial plantations. A question we probe is whether the specific type of inequality that develops depends on which kind of industry people are facing. Mining, for example, is relatively localised in contrast to the other industries, and this has often led to a concentration of wealth from compensation and royalties among a select few officially recognised 'landowners', whereas in logging and agro-industrial developments, which cover vast areas and where compensation is much lower and often only accrues to a few members of the state elite, local inequalities may be less pronounced or take different forms through the impoverishment of the displaced landowners. But, although mines are localised, their impacts need not be: through subcontractors (security firms, dealers in machinery and fencing, transport firms, catering and cleaning companies) and the effects of waste on the environment, the social compass of a mine's effects are considerable.

This volume not only expands the corpus of empirical knowledge on resource extraction and large-scale plantations, but also attempts to reveal connections between these different types of industry, by, for example, showing how plantation projects articulate with other similarly internationally financed projects, like mining. One of the projects observed (PNG Biomass) involves the investment of an oil and gas company (Oil Search) in eucalyptus plantations for the supply of fuel for electricity generation, aimed at satisfying the considerable power requirements of new and existing mining projects. In PNG, at least, this diversification strategy was pioneered by mining company Placer, which built up the country's biggest tree plantation (Healey 1967). By observing linkages between the sectors, we seek to bring attention to the cumulative impact of simultaneous and globally linked capital-intensive projects, and their impact on local life-worlds.

We focus on rural and peri-urban areas of PNG, where the processes leading to the creation of novel forms of inequality associated with capital-intensive projects are more pronounced than in urban areas. We recognise, of course, that rural PNG has never been disconnected from the rest of the national and international economy. After all, there has been a long history of migration from rural to urban areas (or to mining and plantation projects elsewhere), with feedback effects in both directions (Strathern 1975; Levine and Levine 1979; Curry and Koczberski 1998). The Australian colonial government had already promoted labour migrations to plantations, which had repercussions for local social relations (Hayano 1979; Boyd 1981; Ward 1990). At the same time, peri-urban areas, particularly around the larger cities and towns, have become more and more drawn into new social and economic relations (Beer 2017). Nevertheless, peri-urban and rural areas are often transformed at a significantly accelerated pace by the arrival of large-scale, capital-intensive projects, as we will show in this volume.

All contributions to this volume thus offer new perspectives on a continuing engagement with anthropological debates regarding the discipline's place in academic and policy discourses on issues of inequality, and how anthropologists in practice are implicated in the processes of ameliorating or, sometimes, exacerbating these issues. Through a multi-sited (in the case of the chapters on Wampar), comparative and longitudinal ethnography that is informed by long-term fieldwork, the research results suggest further possibilities for refining ethnographic research methods addressing inequalities, especially in their articulation in today's contexts of a global political economy. Methodologically, this volume engages modes intended to bridge gaps in access to knowledge, by trying to involve people in the communities where anthropologists do their studies (in the villages and in local academic institutions) through this open-access publication.

Most of the authors have been working on different topics in their field sites for decades, which gives their contributions the historical depth necessary to identify social change and inequalities. Moreover, their long-term perspective enables a differentiation between economic bubbles (such as the vanilla boom in the East Sepik Province) and long-term economic impacts leading to lasting social inequalities.

Large-Scale Capital and Social Inequality in PNG

Inequality has been a significant theme in the anthropology of New Guinea since the middle of the twentieth century. Earlier studies (Modjeska 1982; Strathern 1982; Godelier 1986; Errington and Gewertz 1987; Strathern 1987; Gewertz and Errington 1999) provide a rich starting point for considering the articulation of significant differences between pre-existing categories and groupings of people (connected to cultural values, gender, gift circulation and production regimes) and the development of various configurations of differentiation (between generations, genders or communities) associated with large-scale mining and other resource extraction projects (Filer 1990; Hyndman 1994a; Bonnell 1999; Bainton 2009; Johnson 2011; Bainton and Macintyre 2013; Gilberthorpe 2013a; Hemer 2013). These changes of configurations have to be compared with changes occurring elsewhere in PNG due to more general trends, ranging from colonialism (Fitzpatrick 1980), the introduction of cash crops and small-scale capitalist enterprises (Epstein 1968; Finney 1973) to the advent of modernity and urbanisation (Levine and Levine 1979). Gewertz and Errington (1998: 345) diagnosed a

> shift in the nature of inequality in Papua New Guinea; a shift whereby differences in life's circumstances and prospects were increasingly understood in class terms. No longer seen as relatively transitory, these differences were shifting from degree to kind, commensurate to incommensurate … Simply put, there was general recognition that an indigenous urban elite was both well established and self-perpetuating, largely as the product of a highly restrictive western-style education …

In PNG, which has many mining ventures as well as oil and gas fields, anthropologists and allied social scientists have been prominent as consultants (Banks 1999; Filer 1999a; Burton 2000; Macintyre 2003; Bainton 2009), as ethnographers of the effects of large-scale projects on local communities (Hyndman 1994a; Jorgensen 1997; Zimmer-Tamakoshi 1997; Kirsch 2006; Bainton 2010; Jacka 2015a) and as a source of reflexive insight into the nature of such encounters (Gerritsen and Macintyre 1991; Filer 1999b; Ballard and Banks 2003; Rumsey and Weiner 2004; Filer and Macintyre 2006; Golub 2007a, 2014; Weiner 2007; Bainton and Macintyre 2013; Gilberthorpe 2013b; Jacka 2018; Bainton 2021). The body of work produced by anthropologists of mining has reflected on environmental impacts, the resistance and transformation of social practices in place prior to projects, the reconfiguration of imagined futures and the lives of workers involved in a local Melanesian setting, whether as labourers (Imbun 2000, 2006; Filer 2021) or executives (Golub and Rhee 2013). Issues relevant to concerns about justice (distribution of risk, costs and benefits, generational and gender inequalities) have been conspicuous in these writings, even when not explicit. A prominent concern for local communities, as well as for non-governmental organisations, governments and corporations sensitive to their international image, has been the environmental damage that has often occurred in areas that the state will not or cannot protect (Hyndman 1994a, 2001; Kirsch 2001, 2007, 2008, 2014; Ballard and Banks 2003; Jacka 2018).

PNG has a long, often fraught, history of mining (Healey 1967; Nelson 1976; Halvaksz 2006, 2008). However, it was only with the local resistance to the Panguna mine on Bougainville, and the civil war to which it led, that social scientists attended to the impact of large-scale resource extraction on local communities in PNG. These events had a profound impact on how resource extraction companies, and the PNG state, seek to mitigate the risk of conflict, with anthropologists prominent in conducting social impact studies (Filer 1999a; Burton 2000; Macintyre 2003; Weiner 2007; Burton et al. 2012). Anthropologists who had previously conducted research among the people later affected by resource extraction (Hyndman 1994a, 1994b; Jorgensen 1997, 2006; Zimmer-Tamakoshi 1997) also began investigating how mines impact on the environment and existing patterns of livelihood, especially after the unprecedented environmental destruction of the Fly River by tailings from the Ok Tedi mine (Banks 2002; Hyndman 2005; Kirsch 2006, 2007, 2008). Others have documented the new economic opportunities mines offer individuals, households and communities, not only as employees (Imbun 2000, 2006; Macintyre 2011; Filer 2021) but also as entrepreneurs or rentiers living off compensation payments (Banks 1996, 1999; Filer 1997; Bainton and Macintyre 2013; Gilberthorpe 2014).

Large-scale mining in PNG has not led to the widespread pauperisation found in Africa or South America, for the state recognises the rights of customary landowners to compensation for the use of their land. During negotiations for the establishment of the Porgera mine in 1988 and 1989, the state instituted the so-called 'development forum', a series of tripartite discussions between the state, the developer and the local landowners to agree on the distribution of benefits from the mine, which has become the norm for further negotiations (Filer 2008). The forced closure of the Panguna mine in 1989 had further strengthened the position of landowners elsewhere in PNG, who have been successfully negotiating for additional benefits in the form of equity positions, trust funds for future generations, preferential employment and business spin-offs (Banks 1996, 2003; Filer 1999b, 2008, 2012a; Gilberthorpe and Banks 2012; Imbun 2013; Golub 2014; Jackson 2015; Banks et al. 2017). There is an inherent paradox in that some of the most detrimental socio-economic changes in mining areas in PNG (e.g. large-scale in-migration, violence and crime, substance abuse, sexual violence) are due to exactly the benefits that flow from the mine to local landowners. This in turn leads the mining companies to invest in 'corporate community development' programs that are inherently conservative in nature, and support 'traditional' institutions (Banks et al. 2013, 2017).

Who counts as a rightful landowner and stands to benefit from these opportunities is highly contested, however. The advent of resource extraction has invariably led to a reframing of local social forms to meet the expectations of the state and companies (Zimmer-Tamakoshi 1997; Jorgensen 2007; Gilberthorpe 2013a; Bacalzo et al. 2014; Minnegal et al. 2015; Minnegal and Dwyer 2017; Skrzypek 2020, 2021), including the invention or 'forging' (Golub 2007b) of unilineal descent groups out of previously fluid cognatic fields (Guddemi 1997; Jorgensen 1997; Ernst 1999; Golub 2007a, 2014; Weiner 2007). Ballard and Banks (2003:297) go so far as to state that 'local communities are only summoned into being or defined as such by the presence or potential presence of a mining project'. This jostling also creates losers between and within communities and regions, losers that are affected by the negative impacts, but receive none of the benefits of large-scale projects. They observe how others gain wealth, while they suffer, and the resentment this generates has already sparked conflicts, as, for example, around the Porgera mine (Jacka 2001, 2015b, 2019; Golub 2021), around the Liquefied Natural Gas (LNG) project in Hela Province (Main and Fletcher 2018; Main 2021a, 2021b), and already much earlier at the Panguna mine on Bougainville—where, as Filer (1990) has shown, compensation flowed to individual titleholders (usually older males) whose failure to redistribute these payments among their kin completely marginalised younger adults, who were then at the forefront of a radical movement that eventually forced the closure of the mine.

Women have also often been politically sidelined in negotiations between landowners and mining companies and have not been able to participate in the new economic opportunities created by the mine to the extent that men have (Beer 2018). Where land was lost to mining or where agriculture was given up entirely, women have further lost social status as they no longer contribute to the household economy and have therefore become more dependent on their men. Most of the negative social impacts of the mine, from large-scale in-migration and alcohol-induced violence to increases in polygyny and extramarital affairs, have disproportionately affected women and children (Hyndman 1994b; Bonnell 1999; Macintyre 2003; Johnson 2011; Menzies and Harley 2012; Hemer 2017; Wardlow 2019, 2020).

But even among those who stand to benefit from large-scale capital-intensive projects, there is a wide diversity of outcomes, partly due to the politics of distributions of these benefits. A study of the financial benefits from the Porgera gold mine (Johnson 2012) has shown that, while payments from the mining company to various stakeholders and the government can be tracked, there is a fundamental lack of data on how that money is further distributed and spent by landowners and the national and sub-national government. Those who reach a position to distribute benefits often stand to disproportionately gain from them, and this development has much to do with the local history of contact and engagement with the extractive company. It is the early brokers, mainly men who are better educated than their peers, who set in motion a development that ends in them—or sometimes their children—holding important positions in the structures set up to distribute benefits (Golub 2014; see also Church, this volume; Dwyer and Minnegal, this volume).

Exactly how the distribution of mining 'benefits' articulates with pre-existing and newly emerging patterns of inequality between local communities and, subsequently, how the distribution and consumption mechanisms initiated by development processes transform and differentiate culturally defined units, therefore remain topics of great analytical significance (Ballard and Banks 2003; Banks 2009, 2019; Golub 2014; Bainton and Owen 2019). Bainton and Macintyre (2016), for example, described how revenue-rich Lihirians spent vast sums on elaborate customary rituals and ceremonial distributions as well as on four-wheel-drive vehicles. For the less cash-fortunate, the surplus from mining was elusive: small business did not lead to mining-based prosperity, and small-scale entrepreneurs then expressed their disappointment in destructive acts such as allowing 'chickens to die, smashing the can crusher and letting several hectares of vegetables to rot in the ground' (Bainton and Macintyre 2013: 156), acts that Bainton and Macintyre characterise as 'ferocious egalitarianism'. As royalties and compensation payments are concentrated in a few 'affected communities' on the main island, this has generated resentment among the other communities who criticise the people living in the 'affected communities' as 'greedy show-offs' (Bainton 2009: 23). Most Lihirians expressed dissatisfaction with social inequality and had to cope with the fact that aspirations for moral equality and material wealth were denied: their anger was mostly directed at the mining company and national government (Bainton 2010). How, why and with what consequences do long-lasting, systematic inequalities in life chances distinguish members of different categories of persons in these settings (Tilly 1998)? Such questions have been difficult to answer for local contexts in PNG.

In contrast to that on mining, the literature on the socio-economic effects of large-scale plantations on local communities in PNG is limited in scope, despite the fact that plantations have a long history in PNG (Lewis 1996). There has been previous research in PNG on plantations dedicated to oil palm (Koczberski and Curry 2005; Koczberski 2007; Koczberski et al. 2012, 2018; Tammisto 2018), coconuts (Panoff 1990), commercial sugar cane (Errington and Gewertz 2004) and (mainly smallholder) coffee (Sexton 1986; West 2012) and cocoa (Curry et al. 2007, 2012). It too has documented various aspects of inequality that develop with the introduction of such plantation schemes. The issue of inequality is much more attenuated in contrast to that existing around mining and oil and gas projects, but it nevertheless arises, and is clearly visible, especially on oil palm or sugar cane plantations, which depend on a clear labour hierarchy between workers, supervisors and managers (Errington and Gewertz 2004; Tammisto 2018). The creation of a whole landscape of plantations, which crowds out previously existing forms of livelihoods, has been likened by Li (2018), in another context in Indonesia, to a form of 'infrastructural violence', as it introduces a series of predatory labour relations and choke points for the capture of rent, creating stark differences in life chances.

In addition to the unequal distribution of benefits from these projects, one of the main dangers lurking behind large-scale plantations is the dispossession of customary landowners due to the large demand for land for these projects. Thus, a lot of recent attention to plantation projects in PNG has focused on the mechanisms known as Special Agricultural Business Leases (SABLs), through which agro-industrial companies have acquired long-term leases over land without the knowledge or consent of most of the local people (Filer 2011a, 2011b, 2012b, 2017; Nelson et al. 2014; Gabriel et al. 2017). Some of these projects were a front to engage in logging, to take out the valuable timber, and then never or only perfunctorily set up oil palm or rubber plantations, leaving behind dusty or water-logged and quickly deteriorating roads, a desolate landscape and shattered dreams of development (Global Witness 2017, 2020; Roberts 2019).

These negative effects are even more pronounced in descriptions of the socio-economic impacts of industrial tree plantations for wood and energy production, which are among the fastest growing monocultures worldwide and have been promoted as carbon sinks. In a review article on the socio-economic effects of industrial tree plantations, Charnley (2005) demonstrates that the establishment of industrial tree plantations often leads to concentration of land ownership, loss of customary access to local resources and socio-economic decline. While large rural landowners and a few plantation employees might benefit, job creation is usually not sufficient for sustainable community development, and is often not available to the more marginalised members of the community (Charnley 2005; Malkamäki et al. 2018). Industrial tree plantations also have a propensity to create a large number of social conflicts. Recent literature reviews on such conflicts (Gerber 2011; Malkamäki et al. 2018), as well as a number of classical and more recent case studies on the establishment of industrial tree plantations (Guha 1990; Peluso 1992; Barney 2004; Gerber et al. 2009; Gerber and Veuthey 2010; Lyons and Westoby 2014; Richards and Lyons 2016), have shown that the majority of these conflicts are due to the displacement of local smallholders and the curtailment of their use of the local ecosystem and thus a significant portion of their livelihood.

This Volume

The five case studies in this volume come from two distinct areas within rural PNG: the mostly rural but increasingly peri-urban Markham Valley in Morobe Province, and the rural and very remote northeastern corner of Western Province. This volume is thus not only a comparison between different forms of large-scale capital-intensive projects, but also a comparison between the two biggest provinces in PNG: Western Province, the largest province by land area; and Morobe Province, the most populous. The two provinces have had rather divergent histories in terms of engagement with large-scale extractive projects, and occupy different positions in the economic landscape of PNG.

Morobe Province saw one of the earliest engagements with large-scale mining in PNG, with the discovery of gold at Edie Creek in the early 1920s, the subsequent gold rush after 1926, and the installation of large-scale dredging operations in the 1930s (Healey 1967; Nelson 1976). After the Second World War, the Markham Valley quickly became the most important area in PNG for cattle ranching (Connell 1979) and mechanised farming, first by expatriate planters, but soon also by local entrepreneurs (Crocombe and Hogbin 1963; Jackson 1965; Fischer 1996; Lütkes 1999). With the completion of the Highlands Highway in 1965, the city of Lae became the gateway to the most densely populated areas of the country, and quickly developed into PNG's most important port and industrial centre. Large-scale extractive industry in the form of mining has only recently made a comeback in Morobe Province, with the development of the Hidden Valley mine and the Wafi-Golpu prospect.

In contrast to Morobe, Western Province was an economic backwater for the whole of the colonial era. Plantation development, so central for the economic life of colonial Papua, remained miniscule in the Western Division (Lewis 1996), and logging only took place on a small scale, with the first foreign-owned large-scale logging companies only becoming interested in the area in the 1980s (Wood 1996). This all changed with the development of the Ok Tedi mine (Hyndman 1994b), and the resulting large-scale environmental catastrophe caused by the pollution of the Fly River (Kirsch 2001). Despite the many years of operation of the Ok Tedi mine, however, Western Province remains one of the most impoverished provinces, with the third-lowest per capita income in PNG (Allen et al. 2005).

Three contributions to this volume also take advantage of a unique contingency: the Wampar in the Markham Valley, an ethnic group that has been studied in depth over decades, is soon to become integral to two large-scale projects: mining and industrial tree plantations (biomass and palm oil). Chapters 2, 3 and 5 refer to long-term research among Wampar and their engagement with large-scale economic projects. Schwoerer, Church and Beer have extensive and recent experience of the communities being drawn into these projects and try to chart the processes whereby social life becomes substantially redefined by its encompassment by the rationales of several TNCs.

The first two chapters, by Schwoerer and Church, focus on claims to ownership of land as one of the central features through which inequalities develop. Land is perennially contested, and with capitalist interests in the mix, conflicts only multiply, which also creates new hierarchies of power and dependence within landowning groups, as both cases show. Tobias Schwoerer shows in Chapter 2 how recent changes in PNG's Land Act and the concurrent development of large-scale industrial tree plantations in the Markham Valley have generated widespread conflicts centred around land that accentuate social, economic and political inequalities within and between social groups among the Wampar. Under current land regulations in PNG, plantation companies can directly engage with customary landowners to access customary land for their purposes. The landowners' decisions to enter into an agreement with one of the two companies competing for land for tree plantations (oil palm and eucalyptus) are shaped by unequal flows of information and existing political alliances. The desire of many Wampar to engage with a company is as much generated by promises of wealth as by the opportunity to secure a legal collective title over their claims to land. Both companies offer to facilitate registrations of Incorporated Land Groups (ILGs) for the customary landowners to access their land. As companies compete, this has led to the duplication of ILG applications within a clan, contrary to the widely circulating notion among government officials that only one ILG can be approved per clan. This form of land registration, in which only parts of a clan are represented in an ILG, threatens to exclude social groups with competing claims, and thus creates novel inequalities, as the first ILG to successfully incorporate and register a title to land tends to set precedents and could dispossess others from rights to land or exclude them from decision-making powers.

Willem Church analyses in Chapter 3 how legal competition around extractive projects can lead to political inequality before such projects begin. By examining three formative periods in the history of the proposed Wafi-Golpu copper/gold mine in Morobe Province, he argues that this competition constitutes a positive feedback process that drives the assembly of politically unequal factions among customary land claimants. The chapter recounts how there was a wide range of possibilities as to who exactly would eventually benefit from Wafi-Golpu when prospecting first began in 1977. However, as cases moved through the courts and communities became settled in the vicinity of the mine, early incumbents became increasingly socially, economically and legally entrenched. In turn, well connected and educated individuals were those best placed to draw together the coalitions necessary to challenge or defend incumbent positions. In the case of Wafi-Golpu, the result of this history has been hierarchical factions, topped by antagonistic members of the local elite, linked to their followers in networks of clientelistic dependence, all perfectly set up for the lopsided distribution of mining benefits. By recounting this case, Church argues that positive feedback in this process explains why the specific beneficiaries of a given project are highly contingent on idiosyncratic historical events, while economic inequality itself is far more robust with respect to local cultural, political and ecological differences across PNG.

Monica Minnegal and Peter Dwyer present a case study from the PNG Liquefied Natural Gas (LNG) project in Chapter 4 and focus on an element already present in the first two chapters: the question of leadership and who ultimately will represent the community in negotiations with project developers. As an extractive industry consolidates in a greenfield, particular men tend to emerge as brokers, acting to negotiate relations between members of their own community and representatives of the state, the companies and neighbouring communities. To the extent that such men are recognised and feted by outsiders, they are vulnerable to becoming complicit in, or submerged by, an ethos of inequality that, initially, they sought to manage on behalf of their constituents. In contexts of these kinds, those men may contribute both to differentiating the domains that they purport to bridge and to enhancing inequalities in their home communities. Such brokers may be powerful, but they are also morally ambiguous individuals—people who cross social boundaries and whose motives and loyalties are thus always open to question. Ultimately, then, these men may experience a personal sense of alienation, failure and loss.

The next two chapters are focused on the effects that anticipation and ideas of a possible future have on people not yet, or only partially, affected by increasing inequalities, and the repercussions this has on ideas of morality. In Chapter 5, Bettina Beer considers changes in the ethical life of the Wampar of the Markham Valley, based on discussions of stinginess and gossip about sharing of food recorded in the 1970s, as well as ethnographic vignettes from fieldwork in 2013 and 2017. She suggests that the social inequalities tending to develop under increasing capital investment and consumerism in the Markham Valley are one reason among others for the changes of values described. The growing importance of money and the desire for consumer goods are implicated in perceived violations of reciprocity, land sales, theft and fraud, as well as in hasty investments in various 'fast money' schemes, or the establishment of risky business ventures. Growing social inequality also leads to new forms of competitive displays of wealth, such as children's birthday parties and fundraising events for school fees (or journeys to sports/church events). As feelings of relative deprivation have spread, and the gap between the desire for goods and the means to get them has steadily widened, discussions of values and the behaviours that they should motivate have become more frequent. These discourses are ubiquitous among relatively impoverished Wampar, but also among the wealthiest, who seek to emphasise the scale of the projects they finance and the concern for the community that motivates them.

Bruce Knauft, in Chapter 6, explores the effects that ripple outwards from areas most directly impacted by mining or petroleum/LNG projects into surrounding areas, and the entrainments of expectation and experienced inequality among peoples not directly impacted by the primary activities of resource extraction. In the Strickland-Bosavi area of PNG, as in many other rural areas in the country, the mere promise of resource development produces cultural dynamics and inequalities that are nonetheless evident. Knauft links the resonating chain of expectations and inequalities that both connect and differentiate areas more or less directly impacted by large-scale resource extraction—its promise, expectation and anticipation. Thereby exposed are the larger dynamics and trajectories of inequality that both connect and polarise peoples who are taken to benefit more, or less, from resource extraction. This throws into relief the illusion that the impact of major resource extraction projects is primarily at the centre of the resource site and its immediately surrounding areas.

In the afterword, Chapter 7, Glenn Banks takes up the themes of this volume and discusses them in relation to his own empirical work on inequalities in the Porgera Valley in the early 1990s. He discerns three central strands running through most of the case studies: the importance of land as a factor in the development of novel inequalities; the issue of leadership and its contested nature in representing the communities affected by resource developments; and the various information flows that structure—but also contest the morality of—inequalities. He combines these reflections with household surveys conducted around the Porgera gold mine in the early to mid-1990s and again in 2019, and finds that persistent inequality is structured around the four axes of geography, gender, hierarchy and residential status.

References

Akin, D. and J. Robbins (eds), 1999. *Money and Modernity: State and Local Currencies in Melanesia*. Pittsburgh: University of Pittsburgh Press.

Allen, B., R.M. Bourke and J. Gibson, 2005. 'Poor Rural Places in Papua New Guinea.' *Asia Pacific Viewpoint* 46(2): 201–217. doi.org/10.1111/j.1467-8373.2005.00274.x

Bacalzo, D., B. Beer and T. Schwoerer, 2014. 'Mining Narratives, the Revival of "Clans" and Other Changes in Wampar Social Imaginaries: A Case Study from Papua New Guinea.' *Le Journal de la Société des Océanistes* 138–139: 63–76. doi.org/10.4000/jso.7128

Bainton, N.A., 2009. 'Keeping the Network Out of View: Mining, Distinctions and Exclusion in Melanesia.' *Oceania* 79(1): 18–33. doi.org/10.1002/j.1834-4461.2009.tb00048.x

———, 2010. *The Lihir Destiny: Cultural Responses to Mining in Melanesia*. Canberra: ANU E Press (Asia-Pacific Environment Monographs). doi.org/10.22459/LD.10.2010

———, 2021. 'Menacing the Mine: Double Asymmetry and Mutual Incomprehension at Lihir.' In N. Bainton, D. McDougall, K. Alexeyeff and J. Cox (eds), *Unequal Lives: Gender, Race and Class in the Western Pacific*. Canberra: ANU Press. doi.org/10.22459/UE.2020.14

Bainton, N.A. and M. Macintyre, 2013. '"My Land, My Work": Business Development and Large-Scale Mining in Papua New Guinea.' In F. McCormack and K. Barclay (eds), *Engaging with Capitalism: Cases from Oceania*. Bingley: Emerald Group Publishing Limited (Research in Economic Anthropology 33). doi.org/10.1108/S0190-1281(2013)0000033008

———, 2016. 'Mortuary Ritual and Mining Riches in Island Melanesia.' In D. Lipset and E.K. Silverman (eds), *Mortuary Dialogues: Death Ritual and the Reproduction of Moral Community in Pacific Modernities*. New York: Berghahn Books. doi.org/10.2307/j.ctvpj7hc4.12

Bainton, N.A., D. McDougall, K. Alexeyeff and J. Cox (eds), 2021. *Unequal Lives: Gender, Race and Class in the Western Pacific*. Canberra: ANU Press. doi.org/10.22459/UE.2020

Bainton, N.A. and J.R. Owen, 2019. 'Zones of Entanglement: Researching Mining Arenas in Melanesia and Beyond.' *The Extractive Industries and Society* 6(3): 767–774. doi.org/10.1016/j.exis.2018.08.012

Ballard, C. and G. Banks, 2003. 'Resource Wars: The Anthropology of Mining.' *Annual Review of Anthropology* 32: 287–313. doi.org/10.1146/annurev.anthro.32.061002.093116

Banks, G., 1996. 'Compensation for Mining: Benefit or Time-Bomb? The Porgera Gold Mine.' In R. Howitt, J. Connell and P. Hirsch (eds), *Resources, Nations and Indigenous Peoples*. Melbourne: Oxford University Press.

———, 1999. 'The Economic Impact of the Mine.' In C. Filer (ed.), *Dilemmas of Development: The Social and Economic Impact of the Porgera Gold Mine, 1989–1994*. Canberra: Asia Pacific Press.

———, 2002. 'Mining and the Environment in Melanesia: Contemporary Debates Reviewed.' *The Contemporary Pacific* 14(1): 39–67. doi.org/10.1353/cp.2002.0002

———, 2003. 'Landowner Equity in Papua New Guinea's Minerals Sector: Review and Policy Issues.' *Natural Resources Forum* 27(3): 223–234. doi.org/10.1111/1477-8947.00057

———, 2009. 'Activities of TNCs in Extractive Industries in Asia and the Pacific: Implications for Development.' *Transnational Corporations* 18(1): 43–59. doi.org/10.18356/b5b9aeca-en

———, 2019. 'Extractive Industries in Melanesia.' In E. Hirsch and W. Rollason (eds), *The Melanesian World*. London; New York: Routledge. doi.org/10.4324/9781315529691-30

Banks, G., D. Kuir-Ayius, D. Kombako and B. Sagir, 2013. 'Conceptualizing Mining Impacts, Livelihoods and Corporate Community Development in Melanesia.' *Community Development Journal* 48(3): 484–500. doi.org/10.1093/cdj/bst025

———, 2017. 'Dissecting Corporate Community Development in the Large-Scale Melanesian Mining Sector.' In C. Filer and P.-Y. Le Meur (eds), *Large-Scale Mines and Local-Level Politics: Between New Caledonia and Papua New Guinea*. Canberra: ANU Press (Asia-Pacific Environment Monographs). doi.org/10.22459/LMLP.10.2017.07

Barney, K., 2004. 'Re-Encountering Resistance: Plantation Activism and Smallholder Production in Thailand and Sarawak, Malaysia.' *Asia Pacific Viewpoint* 45(3): 325–339. doi.org/10.1111/j.1467-8373.2004.t01-1-00244.x

Beer, B., 2017. 'The Intensification of Rural–Urban Networks in the Markham Valley, Papua New Guinea: From Gold-Prospecting to Large-Scale Capitalist Projects.' *Paideuma* 63: 137–158.

————, 2018. 'Gender and Inequality in a Postcolonial Context of Large-Scale Capitalist Projects in the Markham Valley, Papua New Guinea.' *The Australian Journal of Anthropology* 29(3): 348–364. doi.org/10.1111/taja.12298

Benson, P. and S. Kirsch, 2010. 'Capitalism and the Politics of Resignation.' *Current Anthropology* 51(4): 459–486. doi.org/10.1086/653091

Berreman, G.D. and K.M. Zaretsky (eds), 1981. *Social Inequality: Comparative and Developmental Approaches*. New York: Academic Press.

Biersack, A. and J.B. Greenberg, 2006. *Reimagining Political Ecology*. Durham: Duke University Press. doi.org/10.1215/9780822388142

Bonnell, S., 1999. 'Social Change in the Porgera Valley.' In C. Filer (ed.), *Dilemmas of Development: The Social and Economic Impact of the Porgera Gold Mine, 1989–1994*. Canberra: Asia Pacific Press.

Boyd, D., 1981. 'Village Agriculture and Labor Migration: Interrelated Production Activities among the Ilakia Awa of Papua New Guinea.' *American Ethnologist* 8(1): 74–93. doi.org/10.1525/ae.1981.8.1.02a00050

Burawoy, M. (ed.), 2000. *Global Ethnography: Forces, Connections, and Imaginations in a Postmodern World*. Berkeley: University of California Press.

Burton, J., 2000. 'Knowing About Culture: The Handling of Social Issues at Resource Projects in Papua New Guinea.' In A. Hooper (ed.), *Culture and Sustainable Development in the Pacific*. Canberra: Asia Pacific Press.

Burton, J., W. Pondrelei, T. Philipps and R. Lennie, 2012. *Hidden Valley +10: Development and Social Mapping in the Hidden Valley Gold Mine Impact Area, 10 Year Re-Study* (3 volumes). Canberra: ANU Enterprise and Resource Management in Asia-Pacific Program.

Charnley, S., 2005. 'Industrial Plantation Forestry: Do Local Communities Benefit?' *Journal of Sustainable Forestry* 21(4): 35–57. doi.org/10.1300/J091v21n04_04

Connell, J., 1979. 'A Kind of Development? Spatial and Structural Changes in the Beef Cattle Industry of Papua New Guinea.' *GeoJournal* 3(6): 587–598. doi.org/10.1007/BF00186059

Crocombe, R.G. and G.R. Hogbin, 1963. *The Erap Mechanical Farming Project*. Canberra: The Australian National University, Research School of Pacific and Asian Studies, Resource Management in Asia-Pacific Program.

Curry, G.N. and G. Koczberski, 1998. 'Migration and Circulation as a Way of Life for the Wosera Abelam of Papua New Guinea.' *Asia Pacific Viewpoint* 39(1): 29–52. doi.org/10.1111/1467-8373.00052

Curry, G.N., G. Koczberski, J. Lummani, S. Ryan and V. Bue, 2012. 'Earning a Living in PNG: From Subsistence to a Cash Economy.' In M. Robertson (ed.), *Schooling for Sustainable Development: A Focus on Australia, New Zealand, and the Oceanic Region*. Dordrecht: Springer Netherlands. doi.org/10.1007/978-94-007-2882-0_10

Curry, G.N., G. Koczberski, E. Omuru and R.S. Nailina, 2007. *Farming or Foraging? Household Labour and Livelihood Strategies Amongst Smallholder Cocoa Growers in Papua New Guinea*. Perth: Black Swan Press.

Durrenberger, E.P., 2012. *The Anthropological Study of Class and Consciousness*. Boulder: University Press of Colorado.

Epstein, T.S., 1968. *Capitalism, Primitive and Modern: Some Aspects of Tolai Economic Growth*. Canberra: Australian National University Press.

Ernst, T.M., 1999. 'Land, Stories, and Resources: Discourse and Entification in Onabasulu Modernity.' *American Anthropologist* 101(1): 88–97. doi.org/10.1525/aa.1999.101.1.88

Errington, F.K. and D.B. Gewertz, 1987. *Cultural Alternatives and a Feminist Anthropology: An Analysis of Culturally Constructed Gender Interests in Papua New Guinea*. Cambridge; New York: Cambridge University Press.

———, 2004. *Yali's Question: Sugar, Culture, and History*. Chicago: University of Chicago Press.

European Trade Union Institute, 2012. *Benchmarking Working Europe 2012*. Brussels: ETUI-REHS.

Ferguson, J., 2006. *Global Shadows: Africa in the Neoliberal World Order*. Durham: Duke University Press. doi.org/10.1515/9780822387640

Filer, C., 1990. 'The Bougainville Rebellion, the Mining Industry and the Process of Social Disintegration in Papua New Guinea.' *Canberra Anthropology* 13(1): 1–39. doi.org/10.1080/03149099009508487

———, 1997. 'Compensation, Rent and Power in Papua New Guinea.' In S. Toft (ed.), *Compensation for Resource Development in Papua New Guinea*. Port Moresby; Canberra: PNG Law Reform Commission and The Australian National University (Pacific Policy Papers 24).

——— (ed.), 1999a. *Dilemmas of Development: The Social and Economic Impact of the Porgera Gold Mine, 1989–1994*. Canberra: Asia Pacific Press.

———, 1999b. 'The Dialectics of Negation and Negotiation in the Anthropology of Mineral Resource Development in Papua New Guinea.' In A. Cheater (ed.), *The Anthropology of Power: Empowerment and Disempowerment in Changing Structures*. London: Routledge.

———, 2007. 'Local Custom and the Act of Land Group Boundary Maintenance in Papua New Guinea.' In J.F. Weiner and K. Glaskin (eds), *Customary Land Tenure and Registration in Australia and Papua New Guinea: Anthropological Perspectives*. Canberra: ANU E Press (Asia-Pacific Environment Monographs). doi.org/10.22459/CLTRAPNG.06.2007.08

———, 2008. 'Development Forum in Papua New Guinea: Upsides and Downsides.' *Journal of Energy & Natural Resources Law* 26(1): 120–150. doi.org/10.1080/02646811.2008.11435180

———, 2011a. 'The New Land Grab in Papua New Guinea: Case Study from New Ireland Province.' Canberra: State, Society and Governance in Melanesia Program (SSGM Discussion Paper 2011/2).

———, 2011b. 'The Political Construction of a Land Grab in Papua New Guinea.' Canberra: The Australian National University, Crawford School of Economics and Government (READ Pacific Discussion Paper 1).

———, 2012a. 'The Development Forum in Papua New Guinea: Evaluating Outcomes for Local Communities.' In M. Langton and J. Longbottom (eds), *Community Futures, Legal Architecture: Foundations for Indigenous Peoples in the Global Mining Boom*. New York: Routledge.

———, 2012b. 'Why Green Grabs Don't Work in Papua New Guinea.' *Journal of Peasant Studies* 39(2): 599–617. doi.org/10.1080/03066150.2012.665891

———, 2017. 'The Formation of a Land Grab Policy Network in Papua New Guinea.' In S. McDonnell, M.G. Allen and C. Filer (eds), *Kastom, Property and Ideology: Land Transformations in Melanesia*. Canberra: ANU Press. doi.org/10.22459/KPI.03.2017.06

———, 2021. 'Measuring Mobilities and Inequalities in Papua New Guinea's Mining Workforce.' In N. Bainton, D. McDougall, K. Alexeyeff and J. Cox (eds), *Unequal Lives: Gender, Race and Class in the Western Pacific*. Canberra: ANU Press. doi.org/10.22459/UE.2020.13

Filer, C. and M. Macintyre, 2006. 'Grass Roots and Deep Holes: Community Responses to Mining in Melanesia.' *The Contemporary Pacific* 18(2): 215–231. doi.org/10.1353/cp.2006.0012

Finney, B., 1973. *Big-men and Business: Entrepreneurship and Economic Growth in the New Guinea Highlands*. Honolulu: University Press of 'Hawai'i.

Fischer, H. 1996. *Der Haushalt des Darius: Über die Ethnographie von Haushalten* [The Household of Darius: On the Ethnography of Households]. Berlin: Reimer.

Fitzpatrick, P., 1980. *Law and State in Papua New Guinea*. London: Academia Press.

Friedman, J., 1994. *Cultural Identity and Global Process*. London; Thousand Oaks: Sage Publications.

———, 2000. 'Globalization, Class and Culture in Global Systems.' *Journal of World-Systems Research* 6(3): 636–656. doi.org/10.5195/jwsr.2000.198

Friedman, J. and K.E Friedman, 2013. 'Globalization as a Discourse of Hegemonic Crisis: A Global Systemic Analysis.' *American Ethnologist* 40(2): 244–257. doi.org/10.1111/amet.12017

Gabriel, J., P.N. Nelson, C. Filer and M. Wood, 2017. 'Oil Palm Development and Large-Scale Land Acquisitions in Papua New Guinea.' In S. McDonnell, M.G. Allen and C. Filer (eds), *Kastom, Property and Ideology: Land Transformations in Melanesia*. Canberra: ANU Press. doi.org/10.22459/KPI.03.2017.07

Gerber, J.-F., 2011. 'Conflicts Over Industrial Tree Plantations in the South: Who, How and Why?' *Global Environmental Change* 21(1): 165–176. doi.org/10.1016/j.gloenvcha.2010.09.005

Gerber, J.-F. and S. Veuthey, 2010. 'Plantations, Resistance and the Greening of the Agrarian Question in Coastal Ecuador.' *Journal of Agrarian Change* 10(4): 455–481. doi.org/10.1111/j.1471-0366.2010.00265.x

Gerber, J.-F., S. Veuthey and J. Martínez-Alier, 2009. 'Linking Political Ecology with Ecological Economics in Tree Plantation Conflicts in Cameroon and Ecuador.' *Ecological Economics* 68(12): 2885–2889. doi.org/10.1016/j.ecolecon.2009.06.029

Gerritsen, R. and M. Macintyre, 1991. 'Dilemmas of Distribution: The Misima Gold Mine, Papua New Guinea.' In J. Connell and R. Howitt (eds), *Mining and Indigenous Peoples in Australasia*. Sydney: Sydney University Press.

Gewertz, D.B. and F.K. Errington, 1998. 'Sleights of Hand and the Construction of Desire in a Papua New Guinea Modernity.' *The Contemporary Pacific* 10(2): 345–368.

———, 1999. *Emerging Class in Papua New Guinea: The Telling of Difference*. Cambridge; New York: Cambridge University Press. doi.org/10.1017/CBO9780511606120

Gilberthorpe, E., 2013a. 'Community Development in Ok Tedi, Papua New Guinea: The Role of Anthropology in the Extractive Industries.' *Community Development Journal* 48(3): 466–483. doi.org/10.1093/cdj/bst028

———, 2013b. 'In the Shadow of Industry: A Study of Culturization in Papua New Guinea.' *Journal of the Royal Anthropological Institute* 19(2): 261–278. doi.org/10.1111/1467-9655.12032

———, 2014. 'The Money Rain Phenomenon: Papua New Guinea Oil and the Resource Curse.' In E. Gilberthorpe and G. Hilson (eds), *Natural Resource Extraction and Indigenous Livelihoods : Development Challenges in an Era of Globalization*. Farnham: Ashgate.

Gilberthorpe, E. and G. Banks, 2012. 'Development on Whose Terms? CSR Discourse and Social Realities in Papua New Guinea's Extractive Industries Sector.' *Resources Policy* 37(2): 185–193. doi.org/10.1016/j.resourpol.2011.09.005

Global Witness, 2017. 'Stained Trade: How U.S. Imports of Exotic Flooring from China Risk Driving the Theft of Indigenous Land and Deforestation in Papua New Guinea.' Global Witness. Viewed 20 May 2020 at: www.globalwitness.org/documents/19150/stained_trade_310717_lores_pages.pdf

———, 2020. 'Bending the Truth: Purported Rubber Plantation on Papua New Guinea's Manus Island Is a Highly Likely Front for Illegal Logging.' Global Witness. Viewed 20 May 2020 at: www.globalwitness.org/en/campaigns/forests/bending-the-truth/

Godelier, M., 1986. *The Making of Great Men: Male Domination and Power Among the New Guinea Baruya*. Cambridge; New York; Paris: Cambridge University Press; Editions de la maison des sciences de l'homme.

Golub, A., 2007a. 'Ironies of Organization: Landowners, Land Registration, and Papua New Guinea's Mining and Petroleum Industry.' *Human Organization* 66(1): 38–48. doi.org/10.17730/humo.66.1.157563342241q348

———, 2007b. 'From Agency to Agents: Forging Landowner Identities in Porgera.' In J.F. Weiner and K. Glaskin (eds), *Customary Land Tenure and Registration in Australia and Papua New Guinea: Anthropological Perspectives*. Canberra: ANU E Press (Asia-Pacific Environment Monographs). doi.org/10.22459/CLTRAPNG.06.2007.05

———, 2014. *Leviathans at the Gold Mine: Creating Indigenous and Corporate Actors in Papua New Guinea*. Durham: Duke University Press. doi.org/10.1515/9780822377399

———, 2021. '"Restraint without Control:" Law and Order in Porgera and Enga Province, 1950–2015.' In N. Bainton and E. Skrzypek (eds), *The Absent Presence of the State in Large-Scale Resource Extraction Projects*. Canberra: ANU Press (Asia-Pacific Environment Monographs). doi.org/10.22459/AP. 2021.03

Golub, A. and M. Rhee, 2013. 'Traction: The Role of Executives in Localising Global Mining and Petroleum Industries in Papua New Guinea.' *Paideuma* 59: 215–236.

Guddemi, P., 1997. 'Continuities, Contexts, Complexities, and Transformations: Local Land Concepts of a Sepik People Affected by Mining Exploration.' *Anthropological Forum* 7(4): 629–648. doi.org/10.1080/00664677.1997.99 67477

Guha, R., 1990. *The Unquiet Woods: Ecological Change and Peasant Resistance in the Himalaya*. Berkeley: University of California Press.

Halvaksz, J.A., 2006. 'Cannibalistic Imaginaries: Mining the Natural and Social Body in Papua New Guinea.' *The Contemporary Pacific* 18(2): 335–359. doi.org/10.1353/cp.2006.0014

———, 2008. 'Whose Closure? Appearances, Temporality, and Mineral Extraction in Papua New Guinea.' *Journal of the Royal Anthropological Institute* 14(1): 21–37. doi.org/10.1111/j.1467-9655.2007.00476.x

Hann, C.M. and K. Hart, 2011. *Economic Anthropology: History, Ethnography, Critique*. Cambridge, UK ; Malden, MA: Polity Press.

Hayano, D., 1979. 'Male Migrant Labour and Changing Sex Roles in a Papua New Guinea Highlands Society.' *Oceania* 50(1): 37–52. doi.org/10.1002/j.1834-4461.1979.tb01930.x

Healey, A.M., 1967. *Bulolo: A History of the Development of the Bulolo Region, New Guinea*. Canberra: The Australian National University, New Guinea Research Unit.

Hemer, S.R., 2013. *Tracing the Melanesian Person*. Adelaide: University of Adelaide Press.

———, 2017. 'Gender Mainstreaming and Local Politics: Women, Women's Associations and Mining in Lihir.' In C. Filer and P.-Y. Le Meur (eds), *Large-Scale Mines and Local-level Politics: Between New Caledonia and Papua New Guinea*. Canberra: ANU Press (Asia-Pacific Environment Monographs). doi.org/10.22459/LMLP.10.2017.10

Houweling, T.A., A.E. Kunst and J.P. Mackenbach, 2001. 'World Health Report 2000: Inequality Index and Socioeconomic Inequalities in Mortality.' *The Lancet* 357(9269): 1671–1672. doi.org/10.1016/S0140-6736(00)04829-7

Hyndman, D., 1994a. 'A Sacred Mountain of Gold: The Creation of a Mining Resource Frontier in Papua New Guinea.' *The Journal of Pacific History* 29(2): 203–221. doi.org/10.1080/00223349408572772

———, 1994b. *Ancestral Rain Forests and the Mountain of Gold: Indigenous Peoples and Mining in New Guinea.* Boulder: Westview Press.

———, 2001. 'Academic Responsibilities and Representation of the Ok Tedi Crisis in Postcolonial Papua New Guinea.' *The Contemporary Pacific* 13(1): 33–54. doi.org/10.1353/cp.2001.0014

———, 2005. 'Shifting Ecological Imaginaries in the Ok Tedi Mining Crisis in Papua New Guinea.' *Le Journal de la Société des Océanistes* (120–121): 76–93. doi.org/10.4000/jso.396

Imbun, B.Y., 2000. 'Mining Workers or "Opportunist" Tribesmen? A Tribal Workforce in a Papua New Guinea Mine.' *Oceania* 71(2): 129–147. doi.org/10.1002/j.1834-4461.2000.tb02731.x

———, 2006. 'Local Laborers in Papua New Guinea Mining: Attracted or Compelled to Work?' *The Contemporary Pacific* 18(2): 315–333. doi.org/10.1353/cp.2006.0020

———, 2013. 'Maintaining Land Use Agreements in Papua New Guinea Mining: "Business as Usual"?' *Resources Policy* 38(3): 310–319. doi.org/10.1016/j.resourpol.2013.04.003

Jacka, J., 2001. 'On the Outside Looking in: Attitudes and Responses of Non-Landowners Towards Mining at Porgera.' In B.Y. Imbun and P.A. McGavin (eds), *Mining in Papua New Guinea: Analysis & Policy Implications.* Waigani: University of Papua New Guinea Press.

———, 2015a. *Alchemy in the Rain Forest: Politics, Ecology, and Resilience in a New Guinea Mining Area.* Durham: Duke University Press. doi.org/10.1215/9780822375012

———, 2015b. 'Uneven Development in the Papua New Guinea Highlands: Mining, Corporate Social Responsibility, and the "Life Market."' *Focaal* 73: 57–69. doi.org/10.3167/fcl.2015.730105

———, 2018. 'The Anthropology of Mining: The Social and Environmental Impacts of Resource Extraction in the Mineral Age.' *Annual Review of Anthropology* 47: 61–77. doi.org/10.1146/annurev-anthro-102317-050156

———, 2019. 'Resource Conflicts and the Anthropology of the Dark and the Good in Highlands Papua New Guinea.' *The Australian Journal of Anthropology* 30(1): 35–52. doi.org/10.1111/taja.12302

Jackson, G., 1965. *Cattle, Coffee and Land Among the Wain.* Canberra: The Australian National University, New Guinea Research Unit.

Jackson, R.T., 2015. 'The Development and Current State of Landowner Businesses Associated with Resource Projects in Papua New Guinea.' Port Moresby: Papua New Guinea Chamber of Mines and Petroleum.

Jenkins, H. and N. Yakovleva, 2006. 'Corporate Social Responsibility in the Mining Industry: Exploring Trends in Social and Environmental Disclosure.' *Journal of Cleaner Production* 14(3–4): 271–284. doi.org/10.1016/j.jclepro.2004.10.004

Johnson, P., 2011. 'Scoping Project: Social Impact of the Mining Project on Women in the Porgera Area.' Port Moresby: Porgera Environmental Advisory Komiti. Viewed 9 April 2016 at: www.peakpng.org/resources/Women-in-Porgera-Report_Final.pdf (site discontinued).

———, 2012. 'Lode Shedding: A Case Study of the Economic Benefits to the Landowners, the Provincial Government, and the State from the Porgera Gold Mine.' Boroko: National Research Institute (NRI Discussion Paper 124).

Jorgensen, D., 1997. 'Who and What Is a Landowner? Mythology and Marking the Ground in a Papua New Guinea Mining Project.' *Anthropological Forum* 7(4): 599–627. doi.org/10.1080/00664677.1997.9967476

———, 2006. 'Hinterland History: The Ok Tedi Mine and Its Cultural Consequences in Telefolmin.' *The Contemporary Pacific* 18(2): 233–263. doi.org/10.1353/cp.2006.0021

———, 2007. 'Clan-Finding, Clan-Making and the Politics of Identity in a Papua New Guinea Mining Project.' In J.F. Weiner and K. Glaskin (eds), *Customary Land Tenure and Registration in Australia and Papua New Guinea: Anthropological Perspectives.* Canberra: ANU E Press (Asia-Pacific Environment Monographs). doi.org/10.22459/CLTRAPNG.06.2007.04

Keeley, B., 2015. *Income Inequality: The Gap between Rich and Poor.* OECD (OECD Insights). Viewed 18 May 2020 at: www.oecd-ilibrary.org/social-issues-migration-health/income-inequality_9789264246010-en

Kirsch, S., 2001. 'Lost Worlds: Environmental Disaster, "Culture Loss," and the Law.' *Current Anthropology* 42(2): 167–198. doi.org/10.1086/320006

———, 2006. *Reverse Anthropology: Indigenous Analysis of Social and Environmental Relations in New Guinea*. Stanford: Stanford University Press.

———, 2007. 'Indigenous Movements and the Risks of Counterglobalization: Tracking the Campaign Against Papua New Guinea's Ok Tedi Mine.' *American Ethnologist* 34(2): 303–321. doi.org/10.1525/ae.2007.34.2.303

———, 2008. 'Social Relations and the Green Critique of Capitalism in Melanesia.' *American Anthropologist* 110(3): 288–298. doi.org/10.1111/j.1548-1433.2008. 00039.x

———, 2014. *Mining Capitalism: The Relationship Between Corporations and Their Critics*. Berkeley: University of California Press. doi.org/10.1525/ 9780520957596

Knauft, B.M., 1999. *From Primitive to Postcolonial in Melanesia and Anthropology*. Ann Arbor: University of Michigan Press. doi.org/10.3998/mpub.10934

———, 2002. *Exchanging the Past: A Rainforest World of Before and After*. Chicago: University of Chicago Press.

Koczberski, G., 2007. 'Loose Fruit Mamas: Creating Incentives for Smallholder Women in Oil Palm Production in Papua New Guinea.' *World Development* 35(7): 1172–1185. doi.org/10.1016/j.worlddev.2006.10.010

Koczberski, G. and G.N. Curry, 2005. 'Making a Living: Land Pressures and Changing Livelihood Strategies Among Oil Palm Settlers in Papua New Guinea.' *Agricultural Systems* 85(3): 324–339. doi.org/10.1016/j.agsy.2005.06.014

Koczberski, G., G.N. Curry and V. Bue, 2012. 'Oil Palm, Food Security and Adaptation Among Smallholder Households in Papua New Guinea.' *Asia Pacific Viewpoint* 53(3): 288–299. doi.org/10.1111/j.1467-8373.2012.01491.x

Koczberski, G., G.N. Curry, V. Bue, E. Germis, S. Nake and G.M. Tilden, 2018. 'Diffusing Risk and Building Resilience through Innovation: Reciprocal Exchange Relationships, Livelihood Vulnerability and Food Security amongst Smallholder Farmers in Papua New Guinea.' *Human Ecology* 46: 801–814. doi.org/10.1007/s10745-018-0032-9

Levine, H.B. and M.W. Levine, 1979. *Urbanization in Papua New Guinea: A Study of Ambivalent Townsmen*. Cambridge: Cambridge University Press.

Lewis, D.C., 1996. *The Plantation Dream: Developing British New Guinea and Papua 1884–1942*. Canberra: The Journal of Pacific History.

Li, T.M., 2007. *The Will to Improve: Governmentality, Development, and the Practice of Politics*. Durham: Duke University Press. doi.org/10.1515/9780822389781

————, 2018. 'After the Land Grab: Infrastructural Violence and the "Mafia System" in Indonesia's Oil Palm Plantation Zones.' *Geoforum* 96: 328–337. doi.org/10.1016/j.geoforum.2017.10.012

LiPuma, E., 2000. *Encompassing Others: The Magic of Modernity in Melanesia.* Ann Arbor: University of Michigan Press. doi.org/10.3998/mpub.15532

Lütkes, C., 1999. *'Gom': Arbeit und ihre Bedeutung bei den Wampar im Dorf Tararan, Papua-Neuguinea* ['Gom': Work and its Meaning among the Wampar in the Village of Tararan, Papua New Guinea]. Münster; New York: Waxmann.

Lyons, K. and P. Westoby, 2014. 'Carbon Colonialism and the New Land Grab: Plantation Forestry in Uganda and Its Livelihood Impacts.' *Journal of Rural Studies* 36: 13–21. doi.org/10.1016/j.jrurstud.2014.06.002

Macintyre, M., 2003. 'Petztorme Women: Responding to Change in Lihir, Papua New Guinea.' *Oceania* 74(1–2): 120–134. doi.org/10.1002/j.1834-4461.2003.tb02839.x

————, 2011. 'Modernity, Gender and Mining: Experiences from Papua New Guinea.' In K. Lahiri-Dutt (ed.), *Gendering the Field: Towards Sustainable Livelihoods for Mining Communities.* Canberra: ANU E Press. doi.org/10.22459/GF.03.2011.02

Main, M., 2021a. 'The Land of Painted Bones: Warfare, Trauma, and History in Papua New Guinea's Hela Province.' *Anthropological Forum*, 1–19. doi.org/10.1080/00664677.2021.1895070

————, 2021b. 'From Donation to Handout: Resource Wealth and Transformations of Leadership in Huli Politics.' In N. Bainton, D. McDougall, K. Alexeyeff and J. Cox (eds), *Unequal Lives: Gender, Race and Class in the Western Pacific.* Canberra: ANU Press. doi.org/10.22459/UE.2020.12

Main, M. and L. Fletcher, 2018. *On Shaky Ground: PNG LNG and the Consequences of Development Failure.* Sydney: Jubilee Australia Research Centre.

Malkamäki, A., D. D'Amato, N.J. Hogarth, M. Kanninen, R. Pirard, A. Toppinen and W. Zhou, 2018. 'A Systematic Review of the Socio-Economic Impacts of Large-scale Tree Plantations, Worldwide.' *Global Environmental Change* 53: 90–103. doi.org/10.1016/j.gloenvcha.2018.09.001

Menzies, N. and G. Harley, 2012. '"We Want What the Ok Tedi Women Have": Guidance from Papua New Guinea on Women's Engagement in Mining Deals.' Washington DC: The World Bank (J4P Briefing Note No. 7.2). doi.org/10.1596/9780821395066_CH13

Minnegal, M. and P.D. Dwyer, 2017. *Navigating the Future: An Ethnography of Change in Papua New Guinea.* Canberra: ANU Press (Asia-Pacific Environment Monographs). doi.org/10.22459/NTF.06.2017

Minnegal, M., S. Lefort and P.D. Dwyer, 2015. 'Reshaping the Social: A Comparison of Fasu and Kubo-Febi Approaches to Incorporating Land Groups.' *The Asia Pacific Journal of Anthropology* 16(5): 496–513. doi.org/10.1080/14442213. 2015.1085078

Modjeska, N., 1982. Production and Inequality: Perspectives from Central New Guinea. In A. Strathern (ed.), *Inequality in New Guinea Highlands Societies.* Cambridge; New York: Cambridge University Press.

Nelson, H., 1976. *Black, White and Gold: Gold Mining in Papua New Guinea, 1878–1930.* Canberra: Australian National University Press.

Nelson, P.N., J. Gabriel, C. Filer, M. Banabas, J.A. Sayer, G.N. Curry, G. Koczberski and O. Venter, 2014. 'Oil Palm and Deforestation in Papua New Guinea.' *Conservation Letters* 7(3): 188–195. doi.org/10.1111/conl.12058

Panoff, M., 1990. 'Die Rekrutierung von Arbeitskräften für die Plantagen in Neuguinea und ihre demographischen Folgen [The Recruitment of Labour for Plantations in New Guinea and its Demographic Effects].' *Sociologus* 40(2): 121–132.

Payne, K., 2017. *The Broken Ladder: How Inequality Affects the Way We Think, Live, and Die.* New York: Viking.

Peluso, N.L., 1992. *Rich Forests, Poor People: Resource Control and Resistance in Java.* Berkeley: University of California Press. doi.org/10.1525/california/ 9780520073777.001.0001

Richards, C. and K. Lyons, 2016. 'The New Corporate Enclosures: Plantation Forestry, Carbon Markets and the Limits of Financialised Solutions to the Climate Crisis.' *Land Use Policy* 56: 209–216. doi.org/10.1016/j.landusepol. 2016.05.013

Robbins, J. and H. Wardlow (eds), 2005. *The Making of Global and Local Modernities in Melanesia: Humiliation, Transformation and the Nature of Cultural Change.* Aldershot, Burlington: Ashgate.

Roberts, J., 2019. '"We Live Like This": Local Inequalities and Disproportionate Risk in the Context of Extractive Development and Climate Change on New Hanover Island, Papua New Guinea.' *Oceania* 89(1): 68–88. doi.org/10.1002/ ocea.5199

Rondinelli, D.A. and M.A. Berry, 2000. 'Environmental Citizenship in Multinational Corporations: Social Responsibility and Sustainable Development.' *European Management Journal* 18(1): 70–84. doi.org/10.1016/S0263-2373(99)00070-5

Rumsey, A. and J.F. Weiner (eds), 2004. *Mining and Indigenous Lifeworlds in Australia and Papua New Guinea*. Oxon: Sean Kingston Publishing.

Scheper-Hughes, N., 1993. *Death Without Weeping: The Violence of Everyday Life in Brazil*. Berkeley: University of California Press. doi.org/10.1525/9780520911567

Schuerkens, U. (ed.), 2010. *Globalization and Transformations of Social Inequality*. New York: Routledge. doi.org/10.4324/9780203849255

Sexton, L., 1986. *Mothers of Money, Daughters of Coffee: The Wok Meri Movement*. Ann Arbor: UMI Research Press.

Shamir, R., 2010. 'Capitalism, Governance, and Authority: The Case of Corporate Social Responsibility.' *Annual Review of Law and Social Science* 6(1): 531–553. doi.org/10.1146/annurev-lawsocsci-102209-153000

Skrzypek, E., 2020. *Revealing the Invisible Mine: Social Complexities of an Undeveloped Mining Project*. New York: Berghahn.

———, 2021. 'Categorical Dissonance: Experiencing *Gavman* at the Frieda River Project in Papua New Guinea.' In N. Bainton and E. Skrzypek (eds), *The Absent Presence of the State in Large-Scale Resource Extraction Projects*. Canberra: ANU Press (Asia-Pacific Environment Monographs). doi.org/10.2307/j.ctv1zcm2sp.8

Smith, E.A., M. Borgerhoff Mulder, S. Bowles, M. Gurven, T. Hertz and M.K. Shenk, 2010. 'Production Systems, Inheritance, and Inequality in Premodern Societies: Conclusions.' *Current Anthropology* 51(1): 85–94. doi.org/10.1086/649029

Soubbotina, T.P., 2004. *Beyond Economic Growth: An Introduction to Sustainable Development* (2nd edition). Washington, DC: World Bank. doi.org/10.1596/0-8213-5933-9

Strathern, A. (ed.), 1982. *Inequality in New Guinea Highlands Societies*. Cambridge; New York: Cambridge University Press.

Strathern, M., 1975. *No Money on Our Skins: Hagen Migrants in Port Moresby*. Port Moresby and Canberra: The Australian National University, New Guinea Research Unit (New Guinea Research Bulletin 61).

Strathern, M. (ed.), 1987. *Dealing with Inequality: Analysing Gender Relations in Melanesia and Beyond*. Cambridge; New York: Cambridge University Press.

Tammisto, T., 2018. 'Life in the Village Is Free: Socially Reproductive Work and Alienated Labour on an Oil Palm Plantation in Pomio, Papua New Guinea.' *Suomen Anthropologi* 43(4): 19–35.

Tilly, C., 1998. *Durable Inequality*. Berkeley: University of California Press. doi.org/10.1525/9780520924222

———, 2001. 'Relational Origins of Inequality.' *Anthropological Theory* 1(3): 355–372. doi.org/10.1177/14634990122228773

Tsing, A.L., 2005. *Friction: An Ethnography of Global Connection*. Princeton: Princeton University Press. doi.org/10.1515/9781400830596

Ward, G., 1990. 'Contract Labor Recruitment from the Highlands of Papua New Guinea, 1950–1974.' *International Migration Review* 24(2): 273–296. doi.org/10.1177/019791839002400204

Wardlow, H., 2006. *Wayward Women: Sexuality and Agency in a New Guinea Society*. Berkeley: University of California Press. doi.org/10.1525/9780520938977

———, 2019. '"With AIDS I Am Happier Than I Have Ever Been Before."' *The Australian Journal of Anthropology* 30(1): 53–67. doi.org/10.1111/taja.12304

———, 2020. *Fencing in AIDS: Gender, Vulnerability, and Care in Papua New Guinea*. Berkeley: University of California Press. doi.org/10.1525/luminos.94

Weiner, J.F., 2007. 'The Foi Incorporated Land Group: Group Definition and Collective Action in the Kutubu Oil Project Area, Papua New Guinea.' In J.F. Weiner and K. Glaskin (eds), *Customary Land Tenure and Registration in Australia and Papua New Guinea: Anthropological Perspectives*. Canberra: ANU E Press (Asia-Pacific Environment Monographs). doi.org/10.22459/CLTRAPNG.06.2007.07

Welker, M., 2014. *Enacting the Corporation: An American Mining Firm in Post-Authoritarian Indonesia*. Berkeley: University of California Press. doi.org/10.1525/california/9780520282308.001.0001

West, P., 2012. *From Modern Production to Imagined Primitive: The Social World of Coffee from Papua New Guinea*. Durham: Duke University Press. doi.org/10.1515/9780822394846

———, 2016. *Dispossession and the Environment: Rhetoric and Inequality in Papua, New Guinea*. New York: Columbia University Press. doi.org/10.7312/west17878

Wood, M., 1996. 'Logs, Long Socks and the "Tree Leaf" People: An Analysis of a Timber Project in the Western Province of Papua New Guinea.' *Social Analysis* 39: 83–117.

World Bank. 2004. 'Striking a Better Balance: The World Bank Group and Extractive Industries: The Final Report of the Extractive Industries Review. World Bank Group Management Response.' Washington, DC: World Bank.

Zimmer-Tamakoshi, L., 1997. 'When Land Has a Price: Ancestral Gerrymandering and the Resolution of Land Conflicts at Kurumbukare.' *Anthropological Forum* 7(4): 649–666. doi.org/10.1080/00664677.1997.9967478

2

Plantations, Incorporated Land Groups and Emerging Inequalities Among the Wampar of Papua New Guinea

Tobias Schwoerer

Introduction

'If you incorporate your clan as a land group and register your land, you will get a title to that land and become overnight millionaires on your own land!' This enticing promise was made by Benny Allan, the Papua New Guinea Minister for Lands and Physical Planning, at a ceremony on 22 December 2016 when he handed over a land title to representatives of a local land group in a village in Morobe Province. The recipients are members of one of the first groups to have been recognised as an Incorporated Land Group (ILG) and granted a customary land title in the province following the state's recently amended Land Groups Incorporation Act and Land Registration Act. Addressing the crowd gathered in front of the raised platform from where he spoke, he explained that with a customary land title, the landowners could go to the bank and get a loan. Or they could subdivide the area and sublease it to business companies, earning income from rental fees while retaining ownership of the land. Before this legal change, according to the minister, customary land did not have any value. But now, once an ILG has a title to the land, its representatives can go to a registered valuer and have the property appraised, and it will have a value in the millions of kina.[1] Benny Allan then urged the landowners to make use of their land title. He pointed out that for 41 years, since Independence in 1975, customary landowners had just been 'spectators' on their land, but that he had come to empower them now, by handing over the land title to the local ILG.

Whether customary landowners will indeed become millionaires, however, along with the equally pressing question on how any accumulated wealth will be distributed among them, remains to be verified. In this chapter, I argue that changes to the land tenure system in Papua New Guinea (PNG) will, besides igniting conflicts, create new and deepen already existing inequalities based on control over land, access to education and information flows, as well as decision-making power within landowning groups.

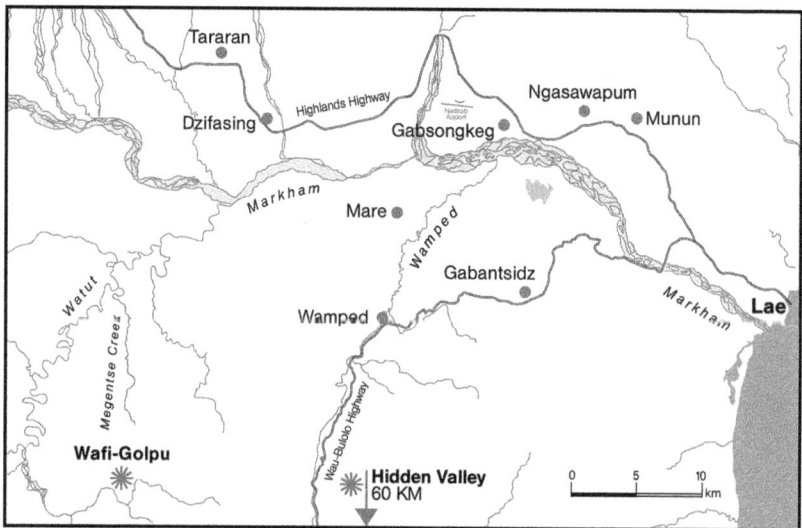

Figure 2.1 Map of Wampar villages in the Markham River Valley.
Source: H. Schnoor.

---

1   One kina is about USD0.30.

My argument is based on recent developments in Dzifasing, a Wampar village in the Markham Valley near the city of Lae, in Morobe Province (see Figure 2.1). There, and in the neighbouring village of Tararan, the development of large-scale industrial tree plantations has generated widespread conflicts, centred around land, that accentuate social, economic and political inequalities within and between local groups. The data for this chapter is based on 12 months of fieldwork between April 2016 and January 2018, in which I observed and participated in interactions between the plantation companies, state representatives and the Wampar, conducted in-depth interviews with Wampar villagers, field officers of the plantation companies, and state officials, and conducted a sample household survey on economic activities and indicators of inequality.

I will first give an overview of the more recent changes in the state law on customary land ownership and then discuss how this law plays out with the entry of two plantation companies, PNG Biomass and New Britain Palm Oil Limited. I will then illustrate through several case studies how new social differentiation by way of ILG formation invites further conflicts, social divisions and inequalities. I will also show that changes in the state law have created a situation where leaders of landowning groups, who often do not have the necessary expertise, capital or political connection to undertake the costly and bureaucratic process to secure their land against rival claimants, eagerly cooperate with the plantation developers who promise to ensure them a title in exchange for a lease on at least a part of their land.

Land Tenure System in PNG

PNG has a dual system of land tenure, wherein traditional, unsurveyed and undocumented land holdings coexist with a modern system of registered and surveyed parcels of alienated and titled land. In the literature, it is often mentioned that 97 per cent of the total land area of PNG is still being held by customary landowners according to customary norms and that only 3 per cent of the land has been alienated by the government, mainly during the colonial administration, and held under freehold or leasehold titles. However, some of this alienated land has been 're-customised' by the disappearance of public records and the redistribution of colonial plantation land after Independence. On the other hand, almost 11 per cent of the total land area had been at least partially alienated through the granting of so-called Special Agricultural Business Leases (SABL) over customary land to private enterprises, mainly between 2008 and 2011 (Filer 2014).

The SABL is a lease–lease-back mechanism, whereby customary landowners can lease their land to the state, who then leases it back either to an ILG—a legal entity representing the customary landowners—a landowner company or even directly to a foreign company if there is consent from the landowners. For the duration of the lease, usually 99 years, all customary rights on the leased land are suspended. In several cases, a few self-declared landowner representatives colluded with local politicians and land department officials to abuse this system and enrich themselves by signing off on leases over vast tracts of land without proper procedures and consultation with the majority of the affected people living on the land. The area was then mostly subleased to foreign-owned logging companies. They often put in a proposal to establish an oil palm plantation or some other agroforestry project as a front to engage in logging (as trees would have to be cut down to plant oil palms), without having any true intentions to establish such a project. Local resistance against these widespread logging operations, coupled with national and international attention, has led to a commission of inquiry into all SABL in 2011, a moratorium on the granting of further SABL, and finally the cancellation of most of them in June 2014 (Filer 2012, 2017; Nelson et al. 2014; Gabriel et al. 2017).

A lot has been written about these SABL and the resulting land grab. Still relatively little attention so far—except by Chand (2017) and Filer (2019)—has been devoted to changes to the Land Groups Incorporation Act and the Land Registration Act that were put in place in 2009 and came into effect in 2012. These changes allowed for a different mechanism for leasing customary land to plantation companies and other businesses. They established a system of Voluntary Customary Land Registration (VCLR), through which customary landowners could apply for a formal collective title to their land, and then lease it out to interested parties.

Before these changes, there were no official titles over customarily held land in PNG. The *Land (Tenure Conversion) Act of 1963* allowed for the transformation of customary (and therefore collectively owned) land into a freehold title that could be held by no more than six customary landowners, but only over smaller portions of land. Until 1979 only about 10,000 hectares have been transformed in this way (Filer 2014: 81).

The *Land Groups Incorporation Act of 1974*, on the other hand, allowed for the registration of landowning groups, establishing them as legal entities, but there was no legislation in place that would grant them a legal title to the land that they held according to their customary norms. The incorporation of land groups only gained traction after changes in the laws regulating the logging industry and the oil and gas industry came into effect, when ILGs became vehicles for disbursing the benefits owed to the local landowners (Weiner 2013; Filer 2014; Minnegal et al. 2015).

With rising pressure especially around cities for secure tenure arrangements, the lands minister at the time, Puka Temu, initiated a policy reform process in 2005 that would allow customary groups to register their land voluntarily, and then be able to lease out that land on their own. He campaigned on a platform of 'turning poor landowners into rich landlords' (Chand 2017: 414). Temu instructed the Director of the National Research Institute to investigate possible amendments to the existing laws regulating land tenure in PNG and created a National Land Development Taskforce that held several consultations with all stakeholders. The suggested changes then passed the parliament in 2009 as the *Land Groups Incorporation (Amendment) Act of 2009* and the *Land Registration (Amendment) Act of 2009*. Both laws took effect in March 2012 and outlined the VCLR process, by which customary landowning groups could register as ILGs and acquire a legal title over their customary land (Chand 2017; Filer 2019).

The VCLR process is an opt-in process, in that a landowning group needs to initiate the process out of their own volition. A landowning group must start the process of incorporation by drawing up a list of all its members, a list of all its landholdings and other rights to resources, indicate the boundaries of the land in question on a sketch map, and draw up a constitution and a governance structure in accordance with the Land Groups Incorporation Act. In drawing up the membership list, the group needs to attach a copy of the birth certificate for each of its members, a provision that was put in place to prevent 'ghost names'.[2] For the people in Dzifasing, this in effect meant that they first had to apply for their birth certificates at the cost of 15 kina per adult. Only a few people were registered at birth with such a document or had to apply for it previously to get a passport. While I was in the field in 2017, the state

---

2   The term 'ghost names' refers to the practice of filling in forms with invented names of non-existing people to create the appearance of an actually existing customary group.

urged individuals over 18 years of age to apply for a national ID card at the same time, as the government had newly implemented a national ID system and wanted to increase the issuing of these ID cards.

The law also states that the group needs to hold a meeting in the area where the members reside and elect a management committee consisting of a chairperson, a vice-chairperson, a secretary, a treasurer and at least two female representatives. The minutes of this meeting, together with the constitution of the Incorporated Land Group, need to be submitted with the application. A general meeting needs to be held every year after that. A quorum in these meetings is reached when at least 60 per cent of all members are present, of which at least 10 per cent need to be of the other gender. Decisions are binding when 60 per cent of the members agree by vote, except in the case of a removal of a member of the management committee, where a majority of 70 per cent of the members present is required.

The sketch map of the land held by the land group needs to indicate the boundaries and any disputed areas along the borders with other neighbouring land groups. The leaders of the disputing adjacent land groups need to sign the map, indicating that they have acknowledged it. If they refuse to do so, the local Ward Councillor or Village Court Magistrate[3] should sign instead. The sketch map, the membership list and all other forms then must be sent to the Department of Lands and Physical Planning, and the submitted information is then forwarded to the District Administrator for verification. After approval, an announcement of incorporation needs to be published in the *National Gazette* and one of the two daily newspapers. After a 30-day objection period, the certificate of incorporation is granted, and the ILG can now start the process to register their pieces of land.

To acquire a customary land title, the ILG will now have to engage a registered surveyor to survey the land, they will need to endorse this registration plan by the Provincial Lands Office, and then make an application to the Director of Customary Land Registration, who forwards a copy of the plan to the Regional Surveyor. A notice needs to be published again in the *National Gazette* and one of the daily newspapers, and an objection period of 90 days follows before the land title then is granted to the ILG. This land title means that the titled land is no

---

3   The Ward Councillor is the local representative on the lowest level of government, (local-level government, or LLG), while the Village Court Magistrate is an official on the lowest level of the PNG court system. PNG has three spheres of government: local, provincial and national, and all three spheres are involved in the process of granting an ILG and a land title.

longer governed by customary law but by state law, and the ILG can now lease out all or portions of the area to anyone and any business entity for a period of up to 99 years. The allodial title, and thus the ownership of the land, remains in perpetuity with the ILG and cannot be transferred.

Some commentators (Chand 2017) are cautiously optimistic about these changes. They point to the design and setup of institutions on all government levels that should prevent the fraudulent uses of ILGs and leases in the past, for example through the use of birth certificates, surveys from certified surveyors and steep punishments for the management committee in case they abuse their trust. Others remain sceptical and point to the lack of information regarding the process of registration, the lack of confidence by landowners in these institutions and the high degree of corruption in the Lands Department as potential downfalls. They ultimately consider Incorporated Land Groups as unsustainable to safeguard the interest of the landowners (Karigawa et al. 2016). Others are deeply concerned by the conceptual changes that take place once customary land ceases to be governed by customary law, as the connection between people and the land is severed:

> the use of ILGs and long-term leases effects a deep, fundamental transformation in the nature of social connection to land and in the nature of power. Made into property, land becomes something to be considered independently of the social world. (Stead 2017a: 80)

Most Wampar I spoke to are less concerned (or not aware) about these more conceptual changes. Most of them struggle to access information, expertise and the capital necessary to start the process of incorporating ILGs and register their land, however. The arrival of plantation developers changed this, as they offered help and encouraged people to engage in the registration process as a prerequisite for the leasing of land to them.

The Arrival of the Plantations

In the Wampar village of Dzifasing, there are currently two companies actively competing to secure land for their planned plantations. They were attracted by the vast and flat expanse of the Markham Valley, covered mostly by savannah grasslands, interspersed with some old-growth forests, mostly along watercourses, and patches of secondary forest of huge rain trees that were introduced with cattle farming in the 1960s and 1970s (see Figure 2.2).

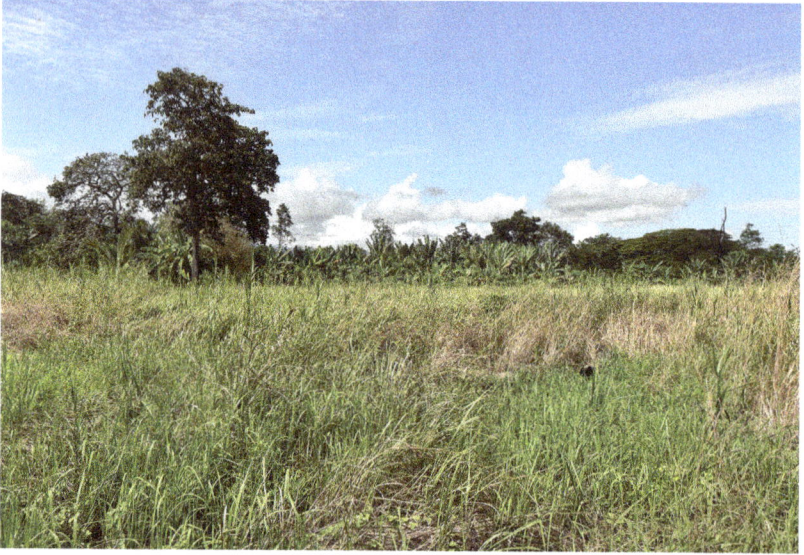

Figure 2.2 Landscape in the Markham Valley.
Source: T. Schwoerer.

Most of the grassland lies idle, as the Wampar only use smaller portions for cocoa and coconut orchards, and gardens where they plant their staples of banana, as well as various fruits and vegetables both for subsistence and for the market in Lae. There are a handful of tractors that are used for mechanised farming, mostly for planting peanuts and watermelons, but only on small patches of usually 1 hectare or less. Much larger parts of the grassland are still used for cattle farming, but both the large-scale collective ranch as well as smaller family-based ranches established in the 1960s, 1970s and 1980s have been run down and stock numbers have been in decline for a while. The mainstay of the local economy had long been the lucrative betelnut trade, and the Wampar were considered as one of the most renowned producers of this mildly narcotic product. When an unidentified pest destroyed most of the betel palm orchards in 2007, the people suddenly had to switch to other economic activities and livelihood strategies, and quite a few were ready to engage with the plantation developers that arrived just a few years later.

Figure 2.3 PNG Biomass eucalyptus plantations.
Source: T. Schwoerer.

The first to arrive was PNG Biomass, a subsidiary of Oil Search Ltd, the largest oil and gas exploration company incorporated in PNG. PNG Biomass plans to establish several eucalyptus plantations on 16,000 hectares of what their website calls 'degraded and underutilised' grassland in the Markham Valley, to supply renewable biomass energy for domestic power consumption. They planned to construct two 15-megawatt wood-fuelled power plants and then start supplying baseload electricity to the Ramu power grid by 2020 (PNG Biomass 2019), but these plans have been delayed, and the company experienced a massive setback in May 2021, when PNG Power cancelled the Power Purchase Agreement with PNG Biomass that it had signed in 2015 (Business Advantage PNG 2021). PNG Biomass has already begun to plant small trial plantations of different eucalyptus and acacia species in 2011, first on 1-hectare patches, later on 2, 5 and 10 hectares, to find out the best-performing species. While I was in the field in 2016 and 2017, they finished their feasibility studies and had started to ramp up the planting of eucalyptus trees. While they had about 350 hectares planted by July 2016, by the end of the year it was already 450 hectares, and by the end of 2017, it was around 2,000 hectares (see Figure 2.3). The trees will be harvested after four to seven years, after which they will re-sprout from the roots for another two to three cycles.

The other company is a relative newcomer to the area: Ramu Agri Industries Limited (RAIL), which in 2008 became a subsidiary of New Britain Palm Oil Limited (NBPOL), a long-standing producer of palm oil in PNG. NBPOL itself had been acquired in 2015 by Sime Darby Plantations, a Malaysian palm oil giant. Officers of RAIL were active during my fieldwork in 2016 and 2017 in contacting and convincing landowners to lease out their customary land to establish oil palm plantations on a total area of at least 5,000 to potentially 10,000 hectares. Ideally, they should be able to acquire at least 5,000 hectares to warrant the construction of a palm oil mill in the area, as trucking the fruit back to the already existing palm oil mill at their Ramu estate would be too expensive.

RAIL offered two different oil palm schemes for landowning groups. The first scheme is mini-estates (of at least 100 hectares or more) that are entirely managed by RAIL, where the landowners will receive a quarterly land rent and a 10 per cent royalty based on the farm-gate price of the harvested fruit, and where some of them could be employed as labourers if they wish. The second is a Village Oil Palm scheme, where landowners participate as out-growers, receive planting materials and fertiliser from the company on credit, and then manage the plantation, provide the labour and sell the oil palm fruit to the company. NBPOL is a signatory to the Roundtable on Sustainable Palm Oil and, according to its website, one of the largest producers of sustainable palm oil worldwide. Being beholden to international standards, NBPOL states that they are committed to gaining free, prior and informed consent from landowners, and that they will use the VCLR system for all new oil palm developments (New Britain Palm Oil Limited 2016; van den Ende and Arihafa 2019).

The two companies both used prominent local leaders as their entry points for their outreach activities to interest others in engaging with their respective plantation projects. This approach meant that the dissemination of information about these projects flowed mostly along already existing lines of kinship and political alliance. PNG Biomass, for example, first established contact through Kelly Jim, a former high school principal who had unsuccessfully run for parliamentary elections in 2007 and 2012. He spread the word about the opportunity to participate in planting trees for money among close kin, affines and political allies, recruiting an initial core group of influential men to commit some of their land to establish trial plantations.

These men first had to convince their brothers and cousins to agree to this venture (or at least to not undertake any measures to block it), as land among the Wampar is collectively held on the level of the extended family or lineage (*mpan* in Wampar). Due to demographic factors and different patterns of inheritance (some apical ancestors divided their land between their sons, others left their landholdings undivided), the size of these lineages varies significantly. Consent could thus be established through anything ranging from an informal discussion between two or three brothers to semi-formal meetings between two dozen or more influential men related in their patriline through a common great-great-grandfather. There are no institutionalised positions of leadership within a lineage. Still, there are usually a few prominent middle-aged men (with adult children) that act as spokespersons of the lineage in dealing with outsiders and other lineages, and a certain degree of respect is usually paid to the first-born son of a first-born son and to the eldest men of each generation.

In a later stage, PNG Biomass also conducted public meetings to attract more landowners, and in the end was able to sign notices of interest with representatives of about 40 different families in Dzifasing and neighbouring Tararan, while others remained sceptical observers. Even some of those who signed the notice of interest were reluctant to be among the first to plant the trees, not knowing what will happen once they do so, nor how much they will actually earn from it. Lack of clear information was often brought up as a major problem in discussions among the people interested in eucalyptus plantations. They wanted clarity regarding the amount of money that they will eventually be paid for the trees, about the potential risks associated with the establishment of these plantations and about the mechanisms through which they will receive money.

One factor for this perceived lack of information was that PNG Biomass had soon organised a core group of 'directors' and 'forum leaders' from among all the lineages that had already agreed to plant eucalyptus. PNG Biomass subsequently focused their dissemination of information through these core leaders. Those leaders then acted as middle-men and controlled the flow of information. Complaints about a lack of clarity were thus not just coming from those who had not yet entered any agreements with the company, but also from those who did and who were not among the key leaders. There was a sense of being left out for many, which they attributed both to the company and their fellow villagers who had become the 'directors' and 'forum leaders'. This situation led to a swirling mass of rumours and expectations, misunderstandings and confusions.

Some villagers thought that the trees were theirs and not owned by the company and that they could do with them as they please. Others received information from the core group of leaders that in addition to the yearly land rent (the amount of which was often not even clear to them when they started planting), they would also eventually be paid per log at harvest time. The prices I heard ranged from a few kina to 50 kina to 500 kina, to even 5,000 kina per log (which would indeed quickly make one a millionaire). With this extreme range of estimates, it was thus no wonder that there were disappointments voiced from those who received their first yearly rental payment of 200–400 kina per hectare, and as it became clear that the harvest was still years away. Prices for the annual rental fee were also lowered over the years, as was the money that was given to the landowners to organise workers in the eucalyptus plantations, such as during the planting of the trees and the cutting of grass. These cuts further fuelled mistrust of the company.

Some of those that remained sceptical or were disappointed by the eucalyptus plantation project were later attracted to the oil palm scheme, which was introduced by the local Member of Parliament (MP), Ross Seymour. Ross is a prominent businessman who grew up in Dzifasing as the son of an Australian father and a Wampar mother. After he won his seat in 2012, he started to look for a company to help landowners turn a profit off their vast land, and at the same time to 'fill up the landscape' to reduce the possibility of settlements of migrants that might be attracted once the prospective Wafi-Golpu gold and copper mine nearby gets underway (see Church, this volume). He directed his public relations officer to contact landowners, and most people interested in oil palm were therefore connected to Ross through ties of kinship and political alliance.

Existing political animosities between the two prominent local leaders and their followers were replicated between the group of supporters of the two plantation companies. Similarly, where ownership or boundary disputes over land already existed, it was almost a rule that one side would support the planting of eucalyptus and the other side oil palm, creating a patchwork of different land use scenarios. Often enough, these different scenarios overlapped, as can be seen in Figure 2.4.

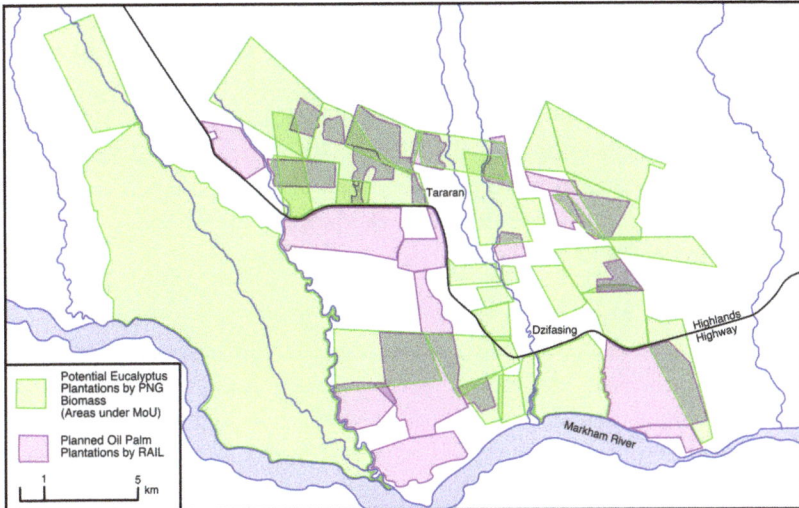

Figure 2.4 Map of prospective plantations.
Source: T. Schwoerer, based on map data in Erias Group (2017) and Ramu Agri
Industries Limited (2018).

Both sides emphasised the potential risk of the other project. Planting oil
palm was suspected to decrease soil fertility, to create a tangle of roots in
which nothing else could be planted and to attract snakes. At the same
time, the planting of eucalyptus was deemed a risky venture, because as
long as the company did not build a power plant, there was no market
for the trees. Most Wampar also worried about the use of herbicides,
and the effect this will have on groundwater quality. Under attack from
the other side, the group of proponents for a project became even more
welded together. The landowner meetings of both groups were veritable
echo chambers, in which they regularly confirmed to each other that they
were on the right track, attempting to shore up support for their project,
discussing positive prospects of their endeavour and the adverse effects
of the opposite project.

As both companies needed a large enough area to warrant the
establishment of plantations, it became imperative for the core supporters
to keep everyone together, and not lose any members to the opposite
project. These attempts were most evident in the eucalyptus project, as
there had already been expressed dissatisfactions and discords among the
growers due to unmet expectations after more than five years. In these
instances, prominent leaders often tried to smooth over disagreements to
keep their allegiance to the company. Landowners in both projects also
regularly maintained a fiction that their ownership and the boundaries

of their land are undisputed, even if there were actual contestations, and they all supported each other's claims in this regard, in order not to alarm the company. This posturing was clearly to present an orderly situation that would also be favourable to the companies. There were also some narratives circulating among the growers that they should be careful in bringing up their disagreements to the company, and not to be too confrontational and demanding, lest the latter abort the whole project, pack up and leave.

With lack of clear information and unable to get verifications through the company officers, most villagers thus evaluated whatever information they could get along the lines of kinship and political alliance, trusting their kin more than others. Decisions were often taken based on very little concrete information, and always involved a high degree of trust. And while the engagement with the companies had an economic motive, people also have a view of their relationship with the company as deeply social, governed by rules of care, trust and reciprocity. At important events, for example, company representatives were given gifts in the form of net bags or other traditional implements (see Figure 2.5). This was done in an attempt to establish deeper reciprocal ties, and company representatives were subsequently asked for contributions to funeral costs if a member of one of the landowning families had died.

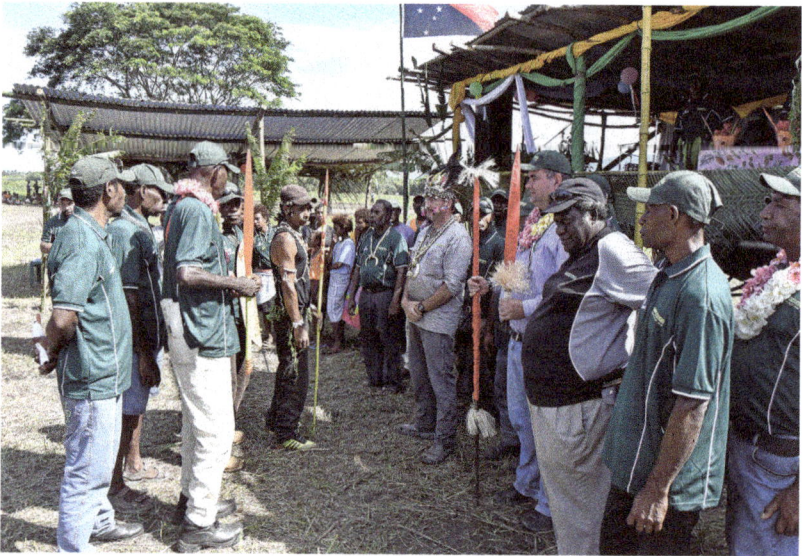

Figure 2.5 Presentation of gifts to representatives of PNG Biomass.
Source: T. Schwoerer.

Motives for engaging the plantation companies are partly economical, as people hope to gain what they perceive as a secure income, even to strike it rich. At the same time, they also see this as a pathway to securing their property claims while staying in control. Thus, some landowners mentioned that they wanted to defend their land from those who do not recognise their rightful ownership. They wanted to receive a secure title to the land. Aside from land disputes with other Wampar lineages, the other commonly raised problem is the increasing settlements of migrants from different areas of PNG. In the current context, one of the best ways for the Wampar to secure land claims and prevent settlements is to sign a contract with the companies that would transform their land into plantations, thus precluding other uses. It was, therefore, no surprise that some of the landowners most eager to engage with the plantation developers were those with the more contentious claims to the land in question (most clear in Figure 2.4, where multiple claims overlap).

With the changes in the land law, it would theoretically be possible to form an ILG and register the area under a formal title through the VCLR mechanism without having to be tied to any of the companies. However, not everyone has sufficient financial resources and the right social networks to facilitate access to the necessary information and key offices that could help navigate the system. Under such constraints, people saw the two companies as the best possible solution.

NBPOL, right from the beginning, has offered help in this regard and has pushed local lineages to get together and form ILGs. The company actively assisted the interested lineages in the bureaucratic procedures, paid for the costs of getting birth certificates and ID cards, and started to survey the parcels of land intended for oil palm plantations. PNG Biomass also publicly stated that they would use ILGs, but only became actively involved in the process of registering ILGs much later into their operations. They focused first on the registration of a landowner business group under the *Business Groups Incorporation Act of 1974*, which would take less time and resources to register than an ILG.

A landowners' business group, as it is composed of landowners as shareholders and directors, can operate on customary land without restrictions, in contrast to a company, which would need a secure title to work on customary land. The business group will in the future oversee the plantation and supply the wood for the power plant operated by PNG Biomass. For the eucalyptus plantations that they had started to establish in 2011, PNG Biomass so far had relied on Clan Land Usage Agreements (CLUA) to gain legal access to customary land.

CLUAs have previously been used by oil palm companies to establish out-grower schemes with traditional landowners, or people who had 'purchased' usage rights to land from the landowners (Koczberski et al. 2013). A CLUA is a contract between the landowners and the company over customary land that is usually not registered or surveyed, without any involvement of the state, and it is, therefore, no real guarantee for the company that the land indeed belongs to the landowners in question, or that the boundaries are accurate. In the case of PNG Biomass, this has then, in turn, led to several disputes over the ownership of the pieces of land on which trees had already been planted, and complaints by members of other lineages that the company had overstepped the boundaries and planted trees on the wrong piece of land.

On several occasions in 2016, the people already growing trees for PNG Biomass called out the company officials for not helping them with registering Incorporated Land Groups and acquiring a formal land title, as NBPOL was doing. By mid-2017, the company had changed its strategy and had started to help lineages interested in eucalyptus plantations to draw up membership lists and fill out the application forms. However, by this time, the VCLR endeavour of PNG Biomass lagged considerably behind the competing effort of NBPOL.

The VCLR System and Property Relations

Implementing the VCLR system among the Wampar was not that straightforward, however. The RAIL officials soon hit a first stumbling block in insisting that ILGs would need to be incorporated on the level of the clan. For many Wampar lineages, this constitutes a problem, as the landholding unit among the Wampar, as mentioned earlier, is not the clan-like formation referred to locally as the *sagaseg*, but the extended family or lineage, or *mpan*. This prominence of the *mpan* goes against the national 'ideology of landownership' in PNG (Filer 2006, 2014: 82), a belief system widely held among state officials and the national elite that ownership of customary land rests exclusively in overarching unilineal descent groups, so either matrilineal or patrilineal clans belonging to one particular village.

The *Land Groups Incorporation (Amendment) Act of 2009* is actually worded to take account of the possibility that different units and not just the clan could be the customary landholding unit. The act defines 'customary landowners' as 'a clan, lineage, family, extended family or other group of persons who hold, or are recognized under custom as holding, rights and interests in customary land, and includes a land group incorporated under the Land Groups Incorporation Act'. But in the rest of the legal text, and an accompanying training manual (GPNG 2012), this use and meaning are inconsistent: in some parts, it is again clearly stated that a customary group can be anything, like a family, extended family, clan or tribe, at other times all these different units are glossed under the name clan. The template forms to be filled out to register an ILG are the most unambiguous indication of this 'landownership' mentality, as they all had the lines '_____ clan of _____ village' as the default option. Company representatives, as well as district and provincial land officers I spoke to, appeared to be unaware of the more open definition of what could constitute a customary landowner in the opinion of the law, as all of them automatically assumed and insisted that ILGs need to be registered on the level of the clan.

Registering as a clan is a significant problem for some *sagaseg* in Dzifasing, as there are deep lines of conflict running through them along lineage lines. Many land conflicts occur between different lineages within the same *sagaseg*. Under the old *Land Group Incorporation Act of 1974*, several *mpan* in Dzifasing had incorporated themselves on the level of the lineage, and not on the level of the *sagaseg*. When the RAIL officials were confronted with this situation, they asked each lineage to draw up a list of its members and a family tree and elect their own executives. They said that they then would collate all the files within one *sagaseg* and arrange a meeting of all executives from all the different lineages of the same *sagaseg* to select the executives for the ILG on the 'clan level'. The majority of lineages not involved in oil palm refused to do so, however, as they did not trust the company or their fellow *sagaseg* members. This refusal then led to the situation that lineages not involved in the oil palm scheme were not automatically included in the ILG to be drawn up in the name of the whole *sagaseg*.

In the end, even the modest endeavour to bring all lineages interested in oil palm under the roof of an ILG foundered in the case of the largest *sagaseg*. Lineage leaders within this *sagaseg* insisted that they wanted to create two ILGs along the lines of the two founding ancestors. The RAIL officials, in the end, accommodated this preference. They began processing an application for two ILGs by amending the registration form, changing 'clan' to 'sub-clan', and named the two ILGs after the *sagaseg* name, followed by the respective eponymous ancestor.

Another difficulty arose due to RAIL and PNG Biomass officials insisting that any person can only be a member of one ILG. Again, the actual law is not conclusive in this regard, as an unamended (and therefore still valid) portion of the *Land Groups Incorporation Act of 1974* includes a provision that 'recognition shall not be refused to a group simply because (a) the members are part only of a customary group or are members of another incorporated land group,' a provision that at least indicates the possibility of a person being a member in several ILGs. This clashes with Schedule 1 of the amended *Land Groups Incorporation (Amendment) Act 2009*, which prescribes that the material to be submitted for an application includes a 'qualification of the group seeking recognition as an incorporated land group stating they are not members of another incorporated land group'. The training manual prepared by the Constitutional and Law Reform Commission is more decisive in this regard and categorically states: 'Multiple membership to ILGs is prohibited, just as a Papua New Guinean belongs to only one clan, tribe or such other land owning social unit' (GPNG 2012: 22). The second part of the statement is of course completely at odds with the actual customary norms of land tenure in many PNG societies, especially in areas with cognatic kinship structures.

Even among the nominally patrilinear Wampar, this provision was challenging to implement, as women are variously seen as enjoying rights both in their natal *sagaseg* and the *sagaseg* of their husband. Women generally retain usage rights to the land of their natal *mpan* even after marriage, and many feared they would now miss out in the distribution of benefits if they are forced to be limited to one group. One lineage leader told me that he had a big argument with his sisters when he had to explain to them that he could not put them on the membership list of his ILG, as they should be listed under the ILG of their respective husbands. Another lineage leader went the opposite direction and respected the wish of his wife and two first-born sons to be registered as members of her brother's ILG, who was from a small *mpan* with large landholdings. At the same time, he listed himself and his third-born son under the ILG of his own rather sizeable *mpan* with only small areas of land. He commented, somewhat bemused, that he now had to break up his family over this whole business.

When PNG Biomass started to assist their group of landowners towards ILG formation, they then duplicated the whole process, as they also registered only those *mpan* interested in planting eucalyptus trees under each *sagaseg* name. This procedure resulted in a duplication of ILGs per *sagaseg* name due to both companies simultaneously registering their respective groups of landowners. RAIL officers told me that they had attempted to reach out to the managers of PNG Biomass, in order to combine their attempts to register the 'clans' into one ILG but that they did not receive any response. Officials in the district administration were also aware of this duplication of applications, and they told me that they wouldn't be able to accept a registration of several ILGs for the same clan and would insist that people sit together and iron out their differences.

Both companies went ahead with sending in their application forms, however. When I left the field in early 2018, there were complaints that both sides attempted to block the other's attempts at registering. As RAIL had the support of the MP with the district administration within his domain, the assumption is that RAIL-facilitated applications will likely pass this stage. At the same time, this hurdle might be difficult for the ILG applications associated with PNG Biomass. Even if the PNG Biomass landowners' ILG applications are sent directly to the national Department of Lands and Physical Planning, they will ultimately also need the approval from the District Administrator. On the other side, some of the prominent individuals engaged in eucalyptus plantations banked on their excellent connections within the Lands Department and predicted that the oil palm–related ILGs would be blocked there.

The process has apparently turned into a race of who can get their ILG approved first. If only one ILG is granted per *sagaseg* name *qua* clan, this might have serious consequences for those *mpan* not included in the successful ILG application. It remains unclear how the disfavoured *mpan* could have an ILG of their own, except maybe by giving themselves another name. Even if there is a possibility for those *mpan* to join the approved ILG later, which means individual members and the corresponding pieces of land would have to be added in the list, they might not be able to pursue their original development plans.

They also would not necessarily have the same (or in fact any) leadership positions within the ILG as they would have if they acquired their own ILG. Joining an already existing ILG is thus not a very attractive proposition for the joining *mpan*, as it could lead to a potential loss of their prior autonomy and power. Once an ILG is registered, the leadership positions on the management committee of the ILG are already filled with members of the *mpan* that initially applied for the ILG. As executives can only be removed by a 70 per cent vote of all ILG members, this means that influential people from the newly entering *mpan* might find it hard to gain similar status. The only way to access these leadership positions is for them to rally enough support to be elected when the seats expire after two years. This outlook of distributing power positions, however, is not particularly rosy, especially if considering the example of another local business. Dzifasing has a large cooperative cattle ranch with a SABL that might still be valid, which was established in the late 1970s after several *mpan* made their land available to develop the ranch and to become shareholders and directors. The current board of directors has now been in power for almost 10 years, as they have postponed continuously the holding of an annual general meeting of all the shareholders where they could be voted out of office.

In terms of property listing for an ILG, I noticed two strategies that people so far used in drawing them up: the first is to only include the unequivocally undisputed land as the property of the *mpan* that will be included in the ILG to minimise the danger that the registration will be objected to. The second strategy is to list all the land properties associated with the same *sagaseg*, including those belonging to the *mpan* that are not involved in forming this particular ILG. The second strategy is to secure the land of the whole *sagaseg* against claims by *mpan* from a different *sagaseg*.

Both cases would make it harder for people that are not included in an ILG to defend their claims in land disputes against other *mpan* that are included in an ILG, as their land is either not included in an ILG, or then covered by an existing ILG of which they are not members. Furthermore, listing property names can itself be a contentious issue, as there is sometimes a dispute whether a name refers to the actual land, a forest, a creek or an old garden and settlement site. This corresponds with different types of resource rights—for example, ownership right over trees, fishing rights in the case of creeks or the usufruct rights to old garden sites. The 'ownership' of these place names is not necessarily disputed, but instead the dispute centres around the question to which category of resource these names belong. The listing of names of properties on a legal form thus cannot be reduced to static types of property that are removed from social life.

As ongoing negotiations continue between the people involved in either the RAIL or PNG Biomass projects, there are some *mpan* who are so far not engaged in either of these projects. Some are reluctant to join an agreement with any company, some want to continue to use their land for cattle, and some have land that was either not identified as a preferred site by the companies (as it is located on the other side of the Markham River, for example) or had been alienated during the colonial period. While these groups are left out of the company projects, it is also more than likely that they will be left behind in the current race for ILG registration and land titling, considering the costs it entails. The fees for applying for birth certificates for all members and publishing notices in the *National Gazette* and a newspaper, and especially the charges to engage a surveyor, are far beyond their means. Without any help and income from the plantation projects, they might thus not be able to secure their land through a formal title.

Surging Tensions and Divisions

With the entry of these large-scale projects, property disputes have surfaced in numbers and varieties of forms beyond the scale of what was known before. The Wampar had always been among the most litigious people in all of Morobe Province when it comes to land disputes, according to some government officials. Some of these land disputes go back generations and had already been brought to the attention of colonial patrol officers. Still, the level of conflicts in 2016 and 2017 was astonishing even for the people themselves. When RAIL officials started a survey of the boundaries with GPS, additional conflicts erupted on the fly, as it was sometimes only during these surveys that neighbouring landowners realised that they did not agree on the exact location of the physical boundaries between their respective parcels of land (see Figure 2.6).

Figure 2.6 Boundary survey for oil palm plantations using GPS equipment.

Source: T. Schwoerer.

These disagreements are not that surprising, as boundaries are not obviously evident on the ground and there are not a lot of occasions where discrepancies in the location of the boundaries become public. In 2017, village leaders thus initiated weekly village meetings in which to discuss these land conflicts in an attempt to mediate. Only a handful of disputes over boundaries or ownership could be solved in these meetings, however. The majority had to be forwarded to mediation by District Land Mediators or then sent directly to the Local (District) Land Court for arbitration. The Local Land Court, however, was already sitting on a massive backlog of existing land disputes and was utterly underfunded. The level of conflict thus has the potential to entrench already existing inequalities, as knowledge about the court system varies widely between the groups, and it disadvantages groups with no connections and less money for legal fees. As Stead (2017b: 375) noted: 'power often goes to those able to translate across ontological difference … those who are best able to position themselves within, and across, both modernist and customary systems of land use and governance'.

These problems with the VCLR process and the tendencies towards entrenching inequalities can also be illustrated in the case of the *sagaseg* that already has a head start and has acquired both an ILG and a title to one piece of customary land in 2016 without the direct help of either of the companies. Several leaders of this group are highly educated, and some even held government positions, and thus had an advantage in terms of information, social networks and resources. They had been present when the then deputy secretary of the Constitutional and Law Reform Commission had visited the local-level government offices and explained the changes in the Land Groups Incorporation Act and the Land Registration Act in 2012. They had befriended him, and a few leaders then made several trips to Port Moresby to have their application submitted in person and to establish good relations with key people in the Department of Lands and Physical Planning. Due to these efforts, other people in Dzifasing constantly referred to the granting of the certificate of incorporation and the customary land title for this *sagaseg* as a 'backdoor deal', fabricated in the corridors of the Lands Department in Port Moresby, and allegedly rushed along with hefty bribes, and they refused to believe in its legality. Nevertheless, it soon became apparent that it was an advantage to have a formal title to land when PNG Biomass started negotiations to use part of this land as the location for their power plant.

There are already tensions growing within this ILG, however, and also with members of other *mpan* within the same *sagaseg* who were not included in the ILG. Within their closest kin network, there were several complaints about the distribution of income from land rent from the eucalyptus project, and about access to cash income from wages from the company. The younger generation especially was at times quite critical of their elders and in one instance, a young man chopped down some eucalyptus trees to show his displeasure at being left out.

In these negotiations, the more vocal and assertive family members have a better chance of being considered, while those who stay in the background tend to be ignored. The lone elder from the first-born branch of this *mpan* is known to be a quiet type and prefers to avoid confrontations with his cousins from subsequent branches who are more competitive and outspoken. There had been some disagreements between him and his cousins before. Still, when the cousins unilaterally decided to transform a particular piece of lineage land into a eucalyptus plantation, his eldest son found a way to get noticed. As he did not dare to appear disrespectful to his classificatory fathers by directly confronting them, he instead went to the village police station to lodge a complaint that his father's family was not consulted in these decisions and did not receive their share of the land rent that was promised them. The police summoned some of the ILG leaders, and the complainants eventually received a share of the land rent, with the caveat that they should not go to the police again and no longer interfere in land matters. The sons thus remained sceptical about whether they will continue to receive their share. This case clearly shows that men (and women) who are not forceful in pressing their claims might easily be overlooked, relegated to the second row, and might lose out in the distribution of the material benefits from these plantation developments.

Another issue concerning this already existing ILG were disputes surrounding its membership. While this ILG encompasses several *mpan* of the same *sagaseg* from three villages, there were still some *mpan* that were left out, or only a few token members were listed. One of the ILG executives explained this with the fact that it costs money to have people registered, as they all needed to get a birth certificate, and that he wanted to get the application submitted as quickly as possible. He assured the other members of the *sagaseg* that they would still be added later once they get their birth certificates. Adding new members is always possible at each annual general meeting, as is, according to their interpretation of the law, the updating of the property list.

The ILG executive remained adamant that he would not include members from a specific *mpan*, however, that he considered to be not really from the same *sagaseg*, as they allegedly are the descendants of a man from a different *sagaseg* that his ancestors had taken under their wing. A leader of this specific *mpan* explained that he would not even think of joining the existing ILG, as he mistrusts the current ILG executives and suspects that they would just embezzle the money paid to the ILG for land leases. He had decided to form his own ILG with the help of RAIL as he had chosen to plant oil palm. He was joined in this endeavour by another *mpan* of the same *sagaseg* consisting only of one older man and his sons. That older man had been listed on the existing ILG together with his properties, while the sons were left out. The sons now feared that with the new laws they would lose the ownership over their lineage land to the ILG executives, as they are not part of that ILG, and they thus decided to form their own ILG and plant oil palm on their land.

The plan of these two *mpan* to establish their common ILG is already facing legal and bureaucratic hurdles, however. When a first application was rejected because they were told that it is not possible to register two ILGs under the same *sagaseg* name, they decided to attempt instead to register under a name that is different from their *sagaseg*. In the case of the second *mpan*, the land that the sons wanted to use for planting oil palm is already listed in the existing ILG. Thus, it is not clear whether their application will be successful, or whether they would first have to ask the executives of the existing ILG to release this property.

In the case of this existing ILG, not only its membership and the distribution of benefits within it has created inequalities, but also the granting of the title over a particular piece of customary land. The boundaries of this specific piece of land are disputed by members of a *mpan* of a different *sagaseg* that own the neighbouring property. One of the leaders of the disputing *mpan* complained that no survey or delineation of the boundary between the two adjoining pieces of land had ever taken place and that the customary land title granted to the ILG also covered parts of his property. He had sent an objection letter both to the granting of the ILG and the granting of the land title during the mandatory objection period, but the Lands Department apparently did not consider his objection.

It appears that the legal requirement that the sketch map accompanying the ILG registration should show any disputed boundaries had not been adhered to, nor the statutory obligation that the Director of Customary Land Registration should verify the survey of land for VCLR in the field. A District Lands Officer confirmed that according to the law the District Administrator would indeed need to approve the property list of each ILG. How this procedure was skipped was a topic filled with conjectures pointing to the politics of the PNG bureaucracy and of winning political offices by presenting oneself as a champion of customary landowners by delivering the first VCLR land titles in Morobe Province before the election.

Such irregularities and skipping of steps in the legal requirements to register customary land are apparently commonplace. Filer (2019: 15–22) has documented, for example, that between 2013 and 2018, 69 of the 87 official notices stating that the Director of Customary Land Registration has received a land investigation report and invites objections within a 30-day period, were published in the *National Gazette* on the same date as the official notices that the director had accepted the survey and will issue a land title over the same portion of land, thus making a mockery of the 30-day objection period.

When members of the disputing *mpan* attempted to delineate the boundaries during a survey of their land intended for an oil palm plantation in 2017, tensions erupted with members of the existing ILG.

The ILG executives insisted that as they already have a legal title to the land showing exactly where the boundaries are between their adjacent land pieces, the claims of the disputing *mpan* must be false. The company thus could lease only the undisputed portion. This conflict left the disputing party with a smaller area of land to lease, plus it would be an uphill battle for them to reclaim what is already titled by the ILG.

If there had been no title, chances for a possibly more even dispute in front of the land mediator or the land courts could exist, as the disputing *mpan* also could rally support through their social networks, including their ties with the MP. As the disputing *mpan* is already involved in two other court battles at the Local Land Court—one in which they claim ownership over a piece of land on which another *mpan* had already planted eucalyptus trees, and another in which their possession of a piece of land where they wanted to plant oil palm had been challenged by two other *mpan* at the same time—they could not afford the costs for another court case. While the outcome of land court matters in PNG is difficult to predict, it becomes clear that the VCLR process in this process has already led to inequalities. The first ILG to successfully gain a title over a piece of land enjoys an advantage, as their opponents first need to raise enough money for legal fees to challenge these titles.

Conclusion

In PNG, property relations are increasingly defined by a land tenure system that tends to simplify a complex set of social relationships expressed in land and resource use rights under the guise of a policy to protect the interests of local populations, culturally broadly referred to as 'customary landowners' and closely linked to the notion of a unilineal 'clan'. This process, however, creates unintended and unimagined repercussions in the lives of the very same social groups the law is meant to serve. Foremost among these effects are the surging of conflicts and the accentuating of both old and new forms of social, economic and political inequalities. The framers of the new land acts operated under the assumption that once land groups are registered, and customary land is titled, there will be cohesive agreements on how the land will be used or leased, and that this will prevent any further disputes over boundaries and ownership of pieces of land. But as I have shown, the implementation of this law, and especially the insistence that ILGs must be incorporated on the level of the 'clan', has stirred up conflicts and competitions that tend to divide more than unify. Even within the already existing ILG, conflicts are ripe about the distribution of the benefits, and people jockey for positions of power and influence.

Among the Wampar in the Markham Valley, the arrival of large-scale plantation projects has created new opportunities to gain wealth. In negotiating this new social terrain of engaging with the companies and the state law, members of Wampar landowning groups draw upon their respective bonds of kinship and political alliance. The unequal access to information and the question of its accuracy and reliability has direct consequences on how people consider options and reach decisions at critical nodes of interaction with the companies, however. And with two plantation companies vying for customary land, and existing lines of conflict over land and politics, this created competing groups that functioned as echo chambers, confirming and strengthening each other in the resolve to continue their engagement with the respective companies.

In a sense, then, the VCLR mechanism is also a way for international capital to access customary land in PNG. With a prevalence of land disputes as among the Wampar, the ability to register an ILG and gain a secure title to land creates a competition between representatives of local landowning groups to outpace each other in a race to be the first to register their ILG and title their land. As the knowledge and costs for doing so exceed the ability and means of most groups, cooperation and collaboration with the plantation companies become necessary. The companies took on parts of the responsibility as it serves their interest, and ILG registration became another platform in their competition to win over as many landowning groups to commit their land to their respective projects. As the companies competed with each other, this led to the duplication of ILGs within a clan, which was contrary to the widely circulating notion among government officials that only one ILG could be approved per clan. This form of land registration, in which only parts of a clan are represented in an ILG, threatens to exclude social groups with competing claims. This will invariably create novel inequalities, as the first ILG to successfully incorporate and achieve a title to land tends to set precedents and could dispossess others from land and political power. The VCLR mechanism and the large-scale plantation projects of the companies are thus connected in a process that creates not only a further social differentiation of local groups but also, more critically, clear winners and losers.

References

Business Advantage PNG, 2021. 'PNG Biomass Green Power Project Under Threat in Papua New Guinea.' *Business Advantage PNG*, 31 May.

Chand, S., 2017. 'Registration and Release of Customary-Land for Private Enterprise: Lessons from Papua New Guinea.' *Land Use Policy* 61: 413–419. doi.org/10.1016/j.landusepol.2016.11.039

Erias Group, 2017. 'PNG Biomass Markham Valley: Environmental Assessment Report.' Viewed 8 January 2022 at: png-data.sprep.org/system/files/PNG-Biomass-EA-Volume1-MainReport.pdf

Filer, C., 2006. 'Custom, Law and Ideology in Papua New Guinea.' *The Asia Pacific Journal of Anthropology* 7(1): 65–84. doi.org/10.1080/14442210600554499

———, 2012. 'Why Green Grabs Don't Work in Papua New Guinea.' *Journal of Peasant Studies* 39(2): 599–617. doi.org/10.1080/03066150.2012.665891

———, 2014. 'The Double Movement of Immovable Property Rights in Papua New Guinea.' *The Journal of Pacific History* 49(1): 76–94. doi.org/10.1080/00223344.2013.876158

———, 2017. 'The Formation of a Land Grab Policy Network in Papua New Guinea.' In S. McDonnell, M.G. Allen and C. Filer (eds), *Kastom, Property and Ideology: Land Transformations in Melanesia*. Canberra: ANU Press. doi.org/10.22459/KPI.03.2017.06

———, 2019. 'Two Steps Forward, Two Steps Back: The Mobilisation of Customary Land in Papua New Guinea.' Canberra: Crawford School of Public Policy, ANU College of Asia & the Pacific (Development Policy Centre Discussion Paper 86). doi.org/10.2139/ssrn.3502585

Gabriel, J., P.N. Nelson, C. Filer and M. Wood, 2017. 'Oil Palm Development and Large-Scale Land Acquisitions in Papua New Guinea.' In S. McDonnell, M.G. Allen and C. Filer (eds), *Kastom, Property and Ideology: Land Transformations in Melanesia*. Canberra: ANU Press. doi.org/10.22459/KPI.03.2017.07

GPNG (Government of Papua New Guinea), 2012. *Training Manual 1 in Implementation of the Land Group Incorporation (Amendment) Act 2009 and the Land Registration (Amendment) Act 2009*. Port Moresby: Constitutional and Law Reform Commission.

Karigawa, L., J. Babarinde and S. Holis, 2016. 'Sustainability of Land Groups in Papua New Guinea.' *Land* 5(2): 14. doi.org/10.3390/land5020014

Koczberski, G., G.N. Curry, D. Rogers, E. Germis and M. Koia, 2013. 'Developing Land-Use Agreements in Commodity Cash Crop Production That Meet the Needs of Landowners and Smallholders.' In G. Hickey (ed.), *Socioeconomic Agricultural Research in Papua New Guinea*. Canberra: Australian Council for International Agricultural Research.

Minnegal, M., S. Lefort and P.D. Dwyer, 2015. 'Reshaping the Social: A Comparison of Fasu and Kubo-Febi Approaches to Incorporating Land Groups.' *The Asia Pacific Journal of Anthropology* 16(5): 496–513. doi.org/10.1080/14442213. 2015.1085078

Nelson, P.N., J. Gabriel, C. Filer, M. Banabas, J.A. Sayer, G.N. Curry, G. Koczberski and O. Venter, 2014. 'Oil Palm and Deforestation in Papua New Guinea.' *Conservation Letters* 7(3): 188–195. doi.org/10.1111/conl.12058

New Britain Palm Oil Limited, 2016. *Sustainability Report 2014/15.* Viewed 8 January 2022 at: www.nbpol.com.pg/wp-content/uploads/downloads/2016/ 11/NBPOL-Sustainability-Report-2015-Final-SP.pdf

PNG Biomass, 2019. 'Project.' Viewed 10 October 2019 at: web.archive.org/ web/20190604010406/http://pngbiomass.com/project/

Ramu Agri Industries Limited, 2018. *Assessment Summaries and Management Plans: Proposed New Plantings by Ramu Agri Industries in Zifasing and Tararan, Morobe Province, Papua New Guinea.* Viewed 8 January 2022 at: rspo.org/public_ consultations/download/dd360bf183de87b

Stead, V., 2017a. *Becoming Landowners: Entanglements of Custom and Modernity in Papua New Guinea and Timor-Leste.* Honolulu: University of Hawai'i Press. doi.org/10.1515/9780824856694

———, 2017b. 'Landownership as Exclusion.' In S. McDonnell, M.G. Allen and C. Filer (eds), *Kastom, Property and Ideology: Land Transformations in Melanesia.* Canberra: ANU Press. doi.org/10.22459/KPI.03.2017.12

van den Ende, S. and A. Arihafa, 2019. 'Local Inclusion in Oil Palm in Papua New Guinea and Solomon Islands.' *ETFRN News* 59: 109–115.

Weiner, J.F., 2013. 'The Incorporated What Group: Ethnographic, Economic and Ideological Perspectives on Customary Land Ownership in Contemporary Papua New Guinea.' *Anthropological Forum* 23(1): 94–106. doi.org/10.1080/ 00664677.2012.736858

3

Factional Competition, Legal Conflict and Emerging Organisational Stratification Around a Prospective Mine in Papua New Guinea[1]

Willem Church

Introduction

Anthropological research on large-scale extractive projects in Papua New Guinea (PNG) has long noted that the benefits and burdens of such projects are unequally distributed among local communities (for example, Filer 1990, 1997; Macintyre 2003; Bainton and Macintyre 2013; Jacka 2015; Banks 2019). In settings as culturally and ecologically diverse as Lihir Island (Bainton and Macintyre 2013) and Porgera Valley in the Highlands of New Guinea (Golub 2014; Jacka 2015), small sections of the local elites disproportionately benefit from contracts, employment and royalties associated with local extractive projects.

1   I would like to thank Bettina Beer, Don Gardner, Tobias Schwoerer, Doris Bacalzo, Sabine Luning and Samuel Whitehead for their valuable feedback on various sections of earlier iterations of this paper. I also acknowledge the participants at the symposia, 'Session on Large-Scale International Capital and Local Inequalities' at the 2018 conference of the Association for Anthropology in Oceania for their useful comments on my presentation of this paper. Finally, I thank the two anonymous reviewers for their helpful feedback on the final draft of this chapter.

These outcomes are familiar for extractive economies suffering from the so-called 'resource curse'.[2] Internationally, it is uncontroversial that high levels of dependency on the extraction of 'point' resources—spatially concentrated resources such as oil wells and mineral deposits—is robustly associated with a range of negative political, economic and social outcomes, including but not limited to relatively low rates of economic growth, fragile political institutions, civil conflict and socio-economic inequality (Gilberthorpe and Papyrakis 2015; Badeeb et al. 2017). Despite this empirical consensus, the precise mechanisms that drive both consistency and variance in these results continue to stimulate debate among academics and policy makers.

In PNG, the entanglement of landowner politics with extractive industries plays a significant part in driving these results. Although the state owns subsoil resources, some 97 per cent[3] of the land in PNG is customarily owned, and landowners gain royalties, preferential employment and contracts from mines (henceforth 'mining-related benefits' (MRBs)). Thus, anthropologists working on resource extraction have stressed for some time that finding 'landowners' and 'impacted communities' of prospective areas is inherently an exclusionary process (Filer 1990, 1997). By demarcating those who are, or are not, landowners and who will, and will not, be impacted, extraction companies and governments selectively dole out the benefits and burdens of resource extraction.

At the same time, even within such demarcated groups, specific individuals profit from extractive projects, while others gain little. It is not necessarily so-called landowners who profit, but those who succeed at being recognised as landowner representatives. Around the Porgera gold mine, Alex Golub explains that:

---

2   The notion of a 'resource curse' grew to prominence following Auty (1993) and Sachs and Warner's (1995) empirical research on the relationship between resource dependency and economic growth. Since these early studies, the scope and disciplines involved in research on the resource curse has expanded considerably (see Badeeb et al. 2017; and Gilberthorpe and Papyrakis 2015 for two recent reviews).

3   At the time of Independence in 1975, 97 per cent of land in PNG was customarily owned. However, there is some reason to question the continuity of this ownership as, by 2011, some 11 per cent of PNG's total land area, all customary land, was under lease–lease-back schemes to government or corporate groups through Special Agricultural Business Leases (SABLs) (see Filer 2011). In theory, in 2016, the O'Neill government 'cancelled' SABLs. However, it is unclear what actual steps the government has taken, as reports of illegal logging continue.

> The benefits of mining have not been distributed equally, and an elite of 'big men' has emerged in Porgera. It is composed of the people appointed to positions of power on the various boards of directors and those who receive lucrative contracts from the mine to provide security, janitorial, and other services. When people speak of 'landowners' it is really these people who they have in mind—large, well-fed men with reputations for prodigality who drive Toyota Land Cruisers with windows tinted to make them opaque. (Golub 2014: 11–12)

The anthropological work to date on social inequality near resource extraction in PNG can be usefully split between accounts of legal and social reconfigurations prior to and in anticipation of the arrival of extractive industries, on the one hand, and studies of the social consequences of novel inequalities once extraction begins, on the other. In the first instance, anthropologists have examined how preparation for extractive industries reconfigured identities and modes of collective action (Jorgensen 1997; Goldman 2007; Jorgensen 2007; Weiner and Glaskin 2007; Weiner 2013; Golub 2014), as well as the speculative, anticipatory character of pending extractive projects (Strathern 1991; Stürzenhofecker 1994; Filer 1997; Dwyer and Minnegal 1998; Minnegal and Dwyer 2017; Skrzypek 2020). These studies connect to a wider literature of millenarian social movements (Worsley 1957; Lawrence 1964; Lindstrom 1993; Jebens 2004; Bainton 2010: 109, 175) and, more recently, 'fast money' schemes (Cox 2018). In contrast, after extraction begins and money begins to flow, anthropologists have examined consequences of novel social inequality, whether changing economies of prestige (Bainton 2010), violent conflict (Filer 1990; Haley and May 2007; Jacka 2015), increasingly exclusionary social relations (Gilberthorpe 2007; Bainton 2009) or the lack of transparency in benefit distribution for both mining and oil extraction (Sagir 2001; Koyama 2004; 2005; Haley and May 2007; Filer 2012).

With the Wafi-Golpu project yet to begin, this chapter fits within the earlier literature and seeks to make one specific contribution. By and large, the flow between the former set of processes (social reconfiguration, anticipation) and the latter (novel social inequalities, violence and so on) tends to be depicted as having an obvious and straightforward source: the state and mine developer desire simple representation on the part of landowners, and therefore those representatives are able to gain a disproportionate share of MRBs. While entirely correct, this fact brackets the question of how, precisely, individuals *become* said representatives, and how this process, in itself, might shape the course of the inequalities to come. Accordingly, this chapter attempts to situate one well-studied dimension of mining in PNG, anticipatory organising and legal registration, squarely within an account of one which has garnered less attention, the competitive relations between different social collectives attempting to gain preferential access to MRBs. In doing so, I hope to illustrate how these legal innovations and competitive engagements drive the creation of social collectives characterised by organisational stratification and bound by promises of clientelistic distributions of benefits.

I attempt to explicitly address this phenomenon by proposing one plausible process, 'stratifying factional competition', that prefigures socio-political relationships for future economic inequality long before extraction commences. Stratifying factional competition, in the case considered in this chapter, involves ongoing legal conflict over MRBs, whereby the increasing financial and social demands for legal disputes require broad, clientelistic coalition building,[4] well-placed contacts and the ability to navigate government bureaucracies. I argue that the skills required for these tasks are unequally distributed across the impacted population, resulting in a narrow elite capable of assembling the necessary coalition, driving the assembly of organisationally stratified, elite-centred factions.

These features first emerge and then are amplified as cases move up the courts, companies build relationships with leaders, the state mandates specific organisational forms for negotiation and communities become settled in the vicinity of the mine. As a consequence, unseating legal incumbents becomes more socially, economically and legally cumbersome. Organising leads to organisations, in which particularly well-educated, well-connected and politically savvy individuals form factions to undertake such competition. *Which* of these factions will ultimately be the beneficiaries of a mine is often the result of highly contingent historical events. My central argument is that the broader organisational demands of factional competition over MRBs robustly drive the emergence of certain *forms* of factions.

I start by briefly introducing the Wafi-Golpu prospect (see Figure 2.1, Chapter 2) and some of the key claimant populations, before summarising the process of MRB distribution in PNG. I then introduce the concept of stratifying factional competition, situating it within the broader literature

---

4   In this chapter, I use 'coalition' in a broad sense of individuals working together towards a given task. When I specifically refer to distinct social entities, such as landowner associations, working together, I use the term 'alliance'.

of factional competition and socio-political stratification. Having introduced the chapter's core model, the second half recounts the three key periods in the history of the Wafi-Golpu project. In the first section, I recount how the social and legal landscape of Wafi-Golpu was cut up in a series of formative court cases in the 1980s. These cases set the boundaries of both land and people while setting the stage with legal incumbents. In the second section, I narrate attempts in the late 1990s and 2000s to shake up this status quo through a dubiously acquired Special Agricultural Business Lease (SABL). Mounting or fending off this legal challenge drove the formation of multiple, competing factions headed by a local, male elite with the education, prestige and coalition-building ability to form such organisations. These challenges cumulated with the introduction of a Special Land Titles Commission (SLTC), designed to resolve land issues of Wafi-Golpu, as discussed in the third and final section. Collectively, I argue that each of these three periods constitute critical junctures in the history of the mine, with critical junctures understood in the conventional sense of 'relatively short periods of time during which there is a substantially heightened probability that agents' choices will affect the outcome of interest' (Capoccia and Kelemen 2007: 348; Pierson 2000; see Schorch and Pascht 2017 for a previous anthropological application of critical junctures to Oceania). These critical junctures, coupled with small initial differences between claimants, has contributed to faction formation, and has resulted in socio-political relations primed for clientelistic, mining-dependent, economic inequality.

This chapter has several limitations: First, the extent that stratifying factional competition, as outlined here, works in a similar fashion in other extractive sites in PNG remains to be established. Wafi-Golpu has a particularly long and tangled legal history, hence my emphasis on courts and legal conflict. However, my sense is that similar processes, even if not as litigious, are present elsewhere.

Second, while based on ethnographic research conducted between 2015 and 2016, the empirical section of the chapter is primarily historical and is based on archival research, legal documents and oral histories of the Wafi-Golpu area. Due to limitations of space and scope, my account is ethnographically light, recounting little of the 'inner workings' of factions, how factions sit within the everyday life they operate in and how factional competition relates to the hopes and dreams of those engaged in such struggles. Instead, this chapter confines itself to critical legal events, and their impact on faction formation and organisational stratification in the Wafi-Golpu area.

Finally, this chapter predominately focuses on the emerging factions among Wampar and Watut-speakers near the Wafi-Golpu project, as this was the central axis of cooperation-*cum*-antagonism that I was able to gather detailed oral and archival histories during my fieldwork in the region over 2016/17. Based on my more limited knowledge of the history of other claimants and interviews with their faction heads, I suspect that stratifying factional competition is just as present within other Wafi-Golpu populations. A more fine-grained history of other claimants may prove this suspicion false.

The Wafi-Golpu Prospect and Claimants

Like most mining projects, the Wafi-Golpu prospect has a history of sale and resale between multiple companies. Conzinc Rio Tinto of Australia Exploration Limited (CRA) discovered and delineated the Wafi gold mineralisation area during the late 1970s and early 1980s. Since this early period, the prospect passed between corporate hands before, from 2008, becoming a joint venture between Newcrest Mining, of Australia and Harmony Gold, of South Africa. At time of writing, the commercial project operates as Wafi-Golpu Joint Ventures (WGJV). The prospect is still technically under exploration, and WGJV submitted its application for a Special Mining Lease in late 2016. If production eventually begins, Wafi-Golpu will be a capital-intensive, relatively labour-sparse operation.[5] There will be no pit, as Wafi-Golpu will be an underground mine requiring kilometres of conveyor belt to transport ore out of the mountain. Exploration drilling is ongoing, but as of April 2018, WGJV estimates the Wafi-Golpu deposit has mineral reserves[6] of 5.5 million ounces of gold and 2.5 million tonnes of copper. The temporal and financial scale of operations will be immense—Wafi-Golpu will have a lifespan of over 25 years and have capital expenditure of some USD5.4 billion. Once built, Wafi-Golpu will be a long-lived, low-production mine—producing

5   According to current estimates, the area above the mine itself will sink into the earth, forming a lake on the mountain. Tailings—a mix of the various chemicals necessary to separate, for example, gold from ore—will be disposed into the Huon Gulf via a pipeline connecting the mine to the coast, in a method known as deep-sea tailings dispersal.
6   Mineral reserves are valuable and legally, economically and technically feasible to extract. Thus, mineral reserves are smaller than mineral resources. WGJV estimates that the Wafi-Golpu deposit has resources of 13 million ounces of gold and 4.4 million tonnes of copper.

an estimated 161,000 tonnes of copper and 266,000 ounces of gold per year—making it what I am told some South African miners call 'a dripping roast'.

Numerous groups claim partial or exclusive customary ownership of the land that will host the project. Understanding the historical formations of factions around Wafi is a task best left for a corkboard covered in names, dates, acronyms and a liberal application of red string linking them together. This complexity is partially a result of the fact that the Wafi-Golpu area sits on a mountain range near two rivers, and at a confluence of linguistic and administrative boundaries.

For purposes of this chapter, it will have to suffice to break the melee into four broad claimant populations:[7] (1) Central Watut–speakers to the west of the prospect, which notably includes the village of Babuaf, (2) Mumeng-speakers of the Bano dialect, including Hengambu, Yanta and Hahiv communities broadly to the south and east, (3) Wampar-speakers, to the more distant northeast, including the Sâb villages of Mare and Wamped,[8] and (4) Piu-speakers, to the immediate south of the Wafe River (see Figure 3.1 for general linguistic boundaries). These names refer to *populations* implicated by Wafi-Golpu, and are by no means political units. Rather, the factions that emerge over the course of this history are drawn from, but are not constituted by, these populations.

Each of the populations mentioned above has some plausible claim to the region where the mine will operate—either through historical occupation, contemporary usage or both. It will suffice to merely note that, like many large-scale projects in PNG, settlement in the region is largely a result of relatively recent migrations, resettlement by Christian missionaries, pacification and the presence of the Wafi-Golpu prospect itself (see Fischer 1963 for Watut history; Ballard 1993a for Yanta and Hengambu history; and Church 2019 for a brief history of pre-mining Wampar involvement in the area). The relatively recent provenance of all contemporary villages and the fact that Wafi-Golpu sits at the confluence of linguistic boundaries makes claims between the different parties particularly intractable. The proximate stake of these claims is recognition as customary landowners of the prospect, and with that recognition, invitation to the Development Forum.

---

7   Each of these names conceals numerous groupings within them, which I will bracket here for purposes of simplicity.

8   Individual names and place names follow Hans Fischer and Bettina Beer's *Wampar–English Dictionary* (2021) to ensure continuity with prior publications.

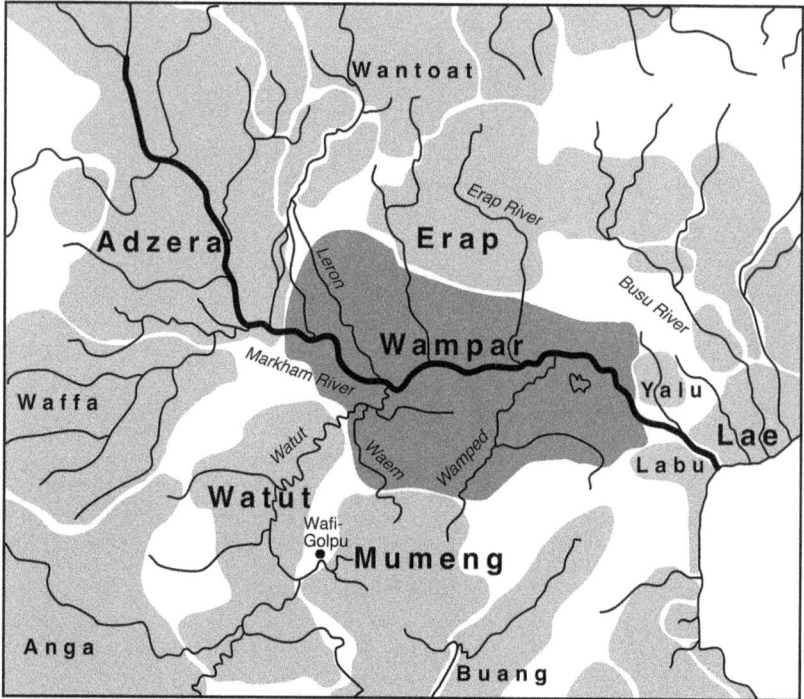

Figure 3.1 Approximate areas of the Wampar and adjacent language groups.
Source: Beer (2006: 108).

The Development Forum

In PNG, companies must acquire a Special Mining Lease (SML) in order to acquire legal permission to develop a large-scale mine. There are multiple stages to apply for an SML, which include submitting an environmental assessment report to the Conservation and Environmental Protection Agency, financial reporting to Treasury, as well as providing the state with the opportunity to buy 30 per cent equity in the project. Perhaps *the* most important stage for impacted communities is the so-called Development Forum.

The Development Forum is a formalised legal mechanism for determining a benefit-sharing agreement of a mine that emerged from the 1988/89 negotiations for the Porgera gold mine (Filer 2012; Golub 2014: 10–11). During this period, the PNG state was engaged in armed civil conflict in Bougainville, sparked by grievances over benefit distribution and environmental damage prompted by the Panguna copper mine, arguably coupled with deeper micronationalist sentiment (see Filer 1990, 1992; Griffin 1990 for a flavour of the debates over the origins of the Bougainville Crisis, also Banks 2005). With these tensions hanging over meetings, the national government created the forum concept to secure support for the Porgera mine. The resulting agreement guaranteed landowners 20 per cent royalties and 5 per cent equity in the project, as well as an obligation for the mine developer to preferentially employ, train and provide business opportunities to local landowners, the affected area and the province, in that order of priority. The Porgera agreements crystalised the Development Forum as the key mechanism for benefit sharing, eventually being legislated into the *Mining Act 1992* (see Filer 2012: 149–51 for an account of the history of the Development Forum in PNG).

More broadly, the Development Forum is part of a tug of war between the national budget in Port Moresby and local interest groups (Filer 1997).[9] For example, in Lihir Gold mines, Lihirians not only gained a 15 per cent equity stake in Lihir Joint Venture, but also secured a 30 per cent transfer of royalties of the mine to local authorities for 'community projects' through the negotiation of the 'integrated benefits package' (Filer 2012: 151). To this end, the evolution of the Development Forum is part of a general pattern of financial resources and responsibility moving away from the national government and towards mining developers, provincial governments, local governments and landowner representatives. Internationally, the Development Forum has been lauded for fostering community participation in the mining sector (MMSD 2002: 211 quoted in Filer 2012: 147). Nevertheless, as the distributional results of other PNG mines illustrate, the extent that such 'participation' is evenly distributed among the local population is questionable.

Legally, the Development Forum includes 'the landholders of the land the subject of the application for the special mining lease and other tenements to which the applicant's proposals relate'.[10] In practice, these landholders are represented by a single landowner association and landowner company

---

9   These trends are part-and-parcel of the worldwide growth of a 'localist' paradigm that has seen resource revenues increasingly redistributed from central to sub-national and local governments, particularly to areas hosting point extractive projects (Arellano-Yanguas 2011: 618, 2017).
10   *Mining Act 1992*, Section 3.

that manages contracts, like catering, security, construction, employment and, in some cases, royalty distribution, 'on behalf of' landowner interests. Thus, the crux of the distributional political economy prior to the Development Forum is: (1) who are customary landowners and, (2) who ought to represent them. Who gets invited to the Development Forum, who heads said companies and who is excluded, is shaped by a series of interactions including litigation in courts, meeting with representatives from the mine developer, and local politicking between would-be landowner representatives. One particular sliver of these interactions concerns this chapter: confrontations between customary claimants at court. My contention is that these confrontations constitute part of a broader process of stratifying factional competition.[11]

Stratifying Factional Competition in Customary Land Litigation

Local-level competitive relations around extractive sites in PNG have been well documented and well discussed within the anthropological literature (Jorgensen 1997, 2007; Gilberthorpe 2007; Weiner 2013; Skrzypek 2020, especially its occasionally violent repercussions; Filer 1990; Ballard and Banks 2003; Banks 2005; Jacka 2015), even if questions of how to examine dissent within and between groups impacted by resource extraction has caused some consternation (see Kirsch 2018; Bainton and Owen 2019 for two contrary views). Given these empirical regularities, and to supplement what has been done so far, there is good reason to reconsider how the now antiqued topic—at least in social anthropology—of factionalism might help to better understand the processes that produce and constitute competition over MRBs.

Factions have not been the focus of anthropological research since the mid-1960s to mid-1970s, when anthropological interest in the topic flourished (Firth 1957; Nicholas 1965; Bailey 1969; Barnes 1969; Gulliver 1971; Kapferer 1976; Silverman and Salisbury 1977; Boissevain 1978). After this brief burst of activity, interest in factionalism declined, even if it continued in adjacent disciplines such as political science.

---

11 Throughout this chapter, I use factional competition and factionalism interchangeably.

While the study of conflict between similarly positioned groups largely faded from social anthropology's purview, social anthropology's sister discipline of archaeology began examining political competition as a means of explaining the emergence of institutionalised inequality (see in particular Brumfiel 1992; Roscoe 1993; Brumfiel and Fox 1994; Hayden 1995; Wiessner 2002; Chacon and Mendoza 2016). For these authors, competition between individuals vying for prestige and wealth becomes a possible driver of socio-political transformation (Wiessner 2002: 234). Such studies differ significantly in their understanding of what factions *are* compared to previous paradigms. The aforementioned anthropological and political science literature focuses on factions as, definitionally, an antagonistic, often maladaptive, subsection of some wider whole. By contrast, the body of research considered here analyses factions by what they *do* in a political environment—social entities, however constructed and constituted, that are engaged in political competition for authority or power over similar aims (see, for example, Brumfiel 1994: 4). Building on this definition, the archaeological literature examines how the classic markers of socio-political change, like the alliances between polities, political centralisation, expanding trade networks and population growth are potentially the consequences and constituent parts of factionalism as individuals attempt to gain the upper hand on their rivals (Brumfiel and Fox 1994: 205).

Drawing inspiration from such studies, this chapter is concerned with how factional competition might drive, within the involved parties, organisational stratification, understood as the extent that there are status differences within a faction, and that those status differences correspond to different degrees of power over the flows of resources within that organisation. As I hope to demonstrate, the extent factions are organisationally stratified near Wafi-Golpu has changed markedly over time. The sets of individuals working together in the 1980s were essentially egalitarian, with particularly vocal individuals able to sway decisions but little more. The landowner associations of the present not only have explicit hierarchies, in the form of a chairman and directors, but these hierarchies also have substantive impacts in decision making and, there is good reason to believe, will shape future benefit distribution. Before laying out my historical narrative, I want to accentuate two features of conflict over MRBs that plausibly aggravate such professionalisation of factions: (1) the social, economic and political requirements for political competition, (2) the progressive increase in the scale of those requirements.

The Costs of Competition

Previous anthropological work on mine-related conflicts has highlighted that politicking around extractive projects takes place at different scales in different spaces (Jorgensen 1997, 2007; Gilberthorpe 2007; Weiner 2013; Skrzypek 2020). This chapter focuses on one, the courtroom, where antagonistic factions seek legal recognition for their claims. However, it is worth situating legal antagonism within the context of coalition building in villages, where would-be leaders struggle for legitimacy and support, and the boardrooms and hotels that host 'consultations', where landowner representatives attempt to extract concessions from the mine developers and state (Golub 2014: 24). Critically, these spaces either require or favour certain resources for participation.

In the courtroom, we have strong *prima facie* grounds for believing that involvement in courts requires, or is at least assisted by, educational, financial, social and political resources.[12] To engage in legal disputes, parties must navigate a range of bureaucratic requirements, such as registering claims and filing briefs. This necessitates literacy and is eased by previous experience with bureaucracies. Courts also place financial demands on participants, requiring legal fees for both courts and lawyers, in addition to the costs of travelling to the court itself. Even for those villages near the mine connected to roads, it can be a solid four-hour drive to Lae. Finally, the economic and logistical difficulties of attending court are not insignificant, especially from villages served only by poor roads or that lack them altogether.

In villages, the Wafi-Golpu prospect is part of a broader universe of future-oriented projects, ranging from proliferating Pentecostal churches to barely veiled Ponzi schemes (à la Cox 2018), all of which promise a better life. Centrally, a would-be faction leader needs to convince others to contribute to their particular cause, and that they are the person to lead. Different faction leaders are more or less successful in building the coalitions necessary for factional competition both by gathering a broad base of supporters and by convincing other prominent individuals to join them, rather than form their own faction. In this way, factions in the

---

12 Legal scholars have long debated the extent that litigation rewards better-off parties through the idea of 'party capability theory' (Galanter 1974; Wheeler et al. 1987; Songer and Sheehan 1992).

Wafi-Golpu region are not dissimilar to competing political candidates, except that instead of the funds and perquisites of the state, the prize is the expected future benefit from the mine.

Finally, state representatives and developer employees have their own ideas about who are legitimate representatives of local landowners. Members of Parliament (MPs) buy certain factions vehicles, enabling them to reach technically 'open' meetings near town otherwise very expensive to reach for the rural population, and official state policy involves winnowing representatives to a single landowner association. These encounters eventually cumulate in the Development Forum, discussed above.

The capacity to meet these different requirements are *not* evenly distributed within claimant populations, favouring certain kinds of people with certain kinds of attributes. All communities in PNG exhibit pre-existing differentiations based on age, gender, education, historical relation to land and experience with wage labour, and are steeped in histories of conflict and cooperation. Specific individuals are systematically favoured to navigate the requirements above: almost exclusively, they are men at the junction of multiple social networks, with experience in government bureaucracy and relatively higher levels of education. As Janet Bujra (1973: 137) argues about factions in general, these patterns are unsurprising: leaders of factions typically come from dominant sectors of society, precisely because they are the ones with resources—in the broadest sense—to recruit large followings and enter political contests.

Positive Feedback and the Escalating Requirements of Competition

Factional competition over MRBs does not occur in a single, decisive round of engagement. Legal conflict over prospect land has a four-decade history, let alone a much deeper history of pre-colonial tensions. In this fashion, there is not one sequence of coalition building, one court case and one consultation that divvies up MRBs. Rather, each 'round' shapes the political landscape of the next. Competition over MRBs is the result of multiple interactions, requiring factional leaders to repeatedly draw together their allies for collective action under increasingly demanding circumstances.

All the features of political conflict considered above become more acute as the mine grows nearer to construction and cases move up the hierarchy of PNG's courts. PNG has a Commonwealth-inspired hierarchy of courts (village, district, national and supreme, each acting as a higher court of appeal) that loosely map onto the levels of government (local, provincial and national). Further, due to widespread customary landownership, the country has a separate hierarchy of courts for dealing with land matters—local land courts, district land courts and provincial land courts—that are legally distinct from the conventional court hierarchy. In practice, the judges that serve on land courts are frequently the same as those that serve on conventional ones.

These nested hierarchies shape factional competition over time. As lower courts make their decisions, the social, economic and legal complications of unseating incumbents rise accordingly. Registering for the Supreme Court is more expensive than for a Local Land Court. As court hearings move further away from the disputed land, attendance requires more elaborate logistical skills and funds. Connections in different locations increase in importance—local contacts in Lae may suffice for the District Land Court, but as cases move to the Supreme Court, contacts and experience in the capital of Port Moresby become more decisive. Finally, court battles in PNG are not straightforward and rarely produce clear, unequivocal outcomes, which only adds to the advantages of detailed knowledge of how the legal system works: such knowledge is an expensive commodity.

Within this context, factional competition exhibits a degree of positive feedback, with early successes improving a faction's ability to conduct future competitions. Whether followers continue to support a particular leader, or the developer or government officials invite groups to stakeholder meetings, depends, in large part, on success in court. These small, public victories are important because as MRBs have not yet begun flowing from the mine, faction leaders face constant problems of credibility. Accordingly, they spend significant amounts of time signalling their moral character and the imminent delivery of MRBs. To this end, legal successes, plans and videos from the mining developer, and appearances of local news broadcasts provide valuable evidence to supporters, while also opening up more opportunities to solicit state agencies and the developers themselves. Escalating costs and feedback loops create a degree of political calcification in the form, although not necessarily the identities involved, in the progressively professionalised factions near the prospect. Given their lower entry costs, early organisation and court successes are more straightforward than later ones. This is not to say upsets are impossible, as I will recount. Rather, such upsets become increasingly difficult as time passes.

Collectively, this ongoing work of opposition and social formation in the shadow of mineral development creates both new identities and organisations, forged by litigating, working and dreaming together. Factional competition demands sufficient unity for coordination for political conflict, and this cooperation contributes to factionalism itself by deepening divisions between those individuals and communities involved. This self-reinforcing loop begins to favour factions that are already recognised as customary landowners, more cohesively coordinated and better funded, and that have gained more credibility and support from the state and the mine developer. As cases advance and time passes, the organisations around Wafi-Golpu, those most centrally placed to control the distribution of benefits, become progressively more centralised, socially calcified and antagonistic as the costs of appeal rises. Consequently, factional competition exhibits 'path dependency', with earlier successes having significantly more downstream consequences than later ones (Arthur 1994; Pierson 2000).[13]

Collectively, these features mean that organisation around mining prospects canalise towards incorporated entities led by someone with a range of 'elite' characteristics—education, contacts and a run of good luck. Through the processes outlined above, the competitive process drives significant power into this person's hands in parallel with factions becoming increasingly legalised. By the time one such leader signs the eventual memorandum of agreement that divides up MRBs, those connected with that faction will likely live a life strikingly different from those who were not so fortunate. Neither elite characteristics nor incorporation, in themselves, explain any subsequent economic inequality; for that, one requires the massive wealth that comes with resource extraction. However, the form of the organisation that the wealth flows through is explicated by the fact that *to have access to those MRBs in the first place,* a specific form of organisation—

---

13 Critically, both the *categories* parties contest over and organise through and the *means* by which they do so also change over time. As has been noted by numerous observers of the extractive industry in PNG, the associations and incorporated groups near mining sites do not straightforwardly reflect local social affiliations (Jorgensen 2007; Weiner 2013). Accordingly, the constant articulation of local political processes and state policy—an articulation based on particular imaginaries of the local—constitutes a process itself. Due to limitations of space and scope, I will not break down this second kind of feedback loop (although see Skrzypek 2020). Here, it necessary to stress that early court cases play a disproportionate role in fixing the salient terms parties compete over later in the process.

patrimonial, elite-dominated and legalised—is systematically favoured by the processes considered above. Having laid out a picture of stratified factional competition, I turn to examining the process through the history of the Wafi-Golpu prospect.

Legal Competition in the Wafi-Golpu Area

The 1980s Cases

At the beginning of the 1980s, there were no rival landowner factions, no prospective mine and no legal incumbents. Precisely because subsequent legal judgements were made with little understanding of local geography and social affiliation, almost anyone living nearby or with some plausible historical connection to the region might have been declared an owner in some capacity. To this end, the historical possibilities of who would be recognised as customary landowners of the Wafi region was extraordinarily open, and the courts might have sliced up social affiliation and land in almost any number of ways.

However, over the course of four cases (see Table 3.1), this space of possibility narrowed significantly and, with it, the odds declined that anybody other than those enshrined in court cases would be recognised as customary landowners. More significantly, the 1980s laid out the names the parties would organise under. Even if appeal overturned earlier decisions, these social divisions would come to shape the Wafi-Golpu region in the years to come.

Table 3.1 Key 1980s court cases pertaining to the Wafi-Golpu area.

Date	Case	Result
6 November 1981	*Babwaf v Engabu.* Local Land Court	'Babwaf clan' awarded ownership of 'Megentse'.
14 May 1982	*Engabu v Babwaf.* District Land Court	1981 decision upheld.
22 March 1984	*Yanta v Engambu, Twangala, Bupu, Omalai, Piu and Perakles*	'Engabu clan' awarded 80 per cent ownership of the land of 'Wafi River Prospect', 'Yanta' awarded 20 per cent.
7 May 1985	*Yanta Clan v Hengabu Clan*	'Hengabu clan' and 'Yanta clan' each awarded 50 per cent ownership of the Wafi River Prospect.

Sources: GPNG (1981, 1982, 1985, 1994).

1981 and 1982

The first pair of cases, the so-called *Megentse* cases, began following a series of fights between men from Babuaf village (to the immediate west of the Wafi prospect, just east of the Watut River) and Hengambu settlers moving into the contemporary sites of Bavaga and Zindaga (near the Waem River). The exact area of Megentse is unclear and has never been demarcated; however, it roughly corresponds to the flat land to the east of the Watut River, going from the Wafe River—a small creek immediately to the south of the prospect (see Figure 2.1, Chapter 2) up until the Waem River (to the immediate north of the prospect).

It is important to stress that, at this initial stage, the land disputes were not about prospective mining. Only in 1977 did Conzinc Rio Tinto of Australia (CRA) Exploration Limited identify the Wafi River to the south of Babuaf as a possible prospect location, while from 1979 to 1981 CRA undertook a series of follow-up studies on Golpu mountain's southern slopes (Ballard 1993a: 32). The *Megentse* cases were not a result of this activity, but were rather prompted by perceived encroachments on land. Had no subsequent mineral exploration taken place, the court cases would likely have been another common, yet largely unremarkable, dispute over customary landownership in the region. It was the subsequent prospect of large-scale extraction that, *ex-post,* gave the cases the importance they have today.

Before turning to the cases, it is necessary to pause and consider the actors who were involved in court, as questions of affiliation became increasingly convoluted following the 1980s cases. Officially, the first pair of cases were between 'Babwaf' [Babuaf] and 'Engabu' [Hengambu] 'clans'. However, neither of these names refer to clans in any sense, nor landholding groups. It is more accurate to see these as the names provided by Australian Patrol Officers *(kiap)* to census units—a different history that will need to be recounted elsewhere.[14] At the time, Babuaf referred to a single, Central Watut–speaking village, while Hengambu referred to a cluster of several settlements that speak the Bano dialect of the Mumeng language.

However, the individuals involved in the case go beyond the occupants of these areas. When the dispute reached the Local Land Court, complainants from Babuaf promptly recruited Wampar-speakers from

---

14   See Ballard (1993a) for a history of Hengambu and Yanta, Ballard (1993b) for a history of Babuaf and Piu, and Fischer (1963) for general Watut history.

Mare and Wamped (see Figure 2.1, Chapter 2) to help testify on their behalf. The men from Babuaf had connections with Wampar-speakers due to their shared history of evangelism; Babuaf was one of the many villages resettled by Wampar evangelists who brought Christianity into the area in the 1930s (Fischer 1963: 235, 2013; also Church 2019). Building on this history, the men from Babuaf enlisted Wampar to assist them in court, the latter relatively well educated due to the Wampar village of Gabsongkeg hosting a mission station since 1911 (Fischer 1992). Wampar elected three people to speak on their behalf[15]—all men, all educated by the Lutheran church and all fluent speakers of Tok Pisin, the lingua franca of PNG.

The Hengambu side involved one particularly knowledgeable man from Hengambu with previous experience as a government official,[16] as well as witnesses from various other Bano-speaking villages. As mentioned above, neither the four witnesses for Babuaf, nor those for Hengambu, constituted anything like 'clans' or landowner associations. Rather they are better understood as 'action-sets', 'ad-hoc unit[s] for collective action', ephemeral and gathered for the specific purpose of testimony in court (Gulliver 1971: 18). To the extent that the parties constituted conflicting social entities over authority and power over a certain strip of land, they were factions in the understanding of this chapter, albeit fleeting ones.

At the court itself, the three Wampar witnesses recounted the late nineteenth-century Wampar history of migration from the disputed area, claiming it as their land by right of ancestral occupation. The Wampar witnesses also generously include Watut-speakers at various points of their story, claiming to have co-resided in historical villages in the region and speaking one local language (GPNG 1981). The sole Babuaf witness finished his testimony with: 'Because of this, Wampar and Babwaf know that the land belongs to Wampar and Babwaf.'[17] Hengambu witnesses, in turn, recounted occasional fights with Wampar up at the river Waem but argued that the land was mostly vacant when they arrived, with Watut-speakers confined to the west of the Watut River (GPNG 1981).

---

15   Peats Go, Intu Ninits and Gau Monz (see Church 2019 for Peats Go's biography). The research project this chapter is part of observes a mixed naming strategy, using pseudonyms where possible for living individuals and real names for historical ones. For living individuals that are prolific public figures, we use their real names.

16   Kitumbing Nganiatuk, who also was Chris Ballard's key informant for his account of Hengambu history in his social mapping study of the Wafi-Golpu area (Ballard 1993a).

17   Esera Kwako (GPNG 1981).

The Local Land Court awarded the case to 'Babwaf clan' owing to the lack of Hengambu witnesses. Hengambu representatives promptly appealed to the Provincial Land Court, which upheld the earlier decision.

From today's vantage point, confusions of geography and affiliation muddle both the testimonies as well as the summary of the decision itself—part of a broader trend in the Wafi cases—and the historic merging of the Watut and Wampar-speaking parties has driven significant confusion in subsequent cases. Wampar testimonies were likely critical for the court victory, given their substantial testimony and linguistic proficiency. However, whatever these advantages, Wampar were not inscribed as one of the parties of the case, a fact that would haunt them in years to come. Regardless, these first legal exchanges entered Hengambu and Babuaf into the jurisprudential annuals of PNG, while the cases would further entangle Babuaf and Wampar in the years to come. More fundamentally, the 1982 case resulted in two new legal categories with ambiguous membership criteria, Hengambu and Babuaf clans, with the latter as customary landowners of Megentse.

1984 and 1985

As the *Megentse* case was contested in the courts, the Wafi prospect owners approached Yanta peoples in the village of Venembeli to assist with the nascent prospecting. Yanta, like Hengambu, are Bano dialect Mumeng-speakers, and like both Hengambu and Babuaf, the name 'Yanta' has its origin in *kiap* census groupings (Ballard 1993a). By the time CRA began their search for gold, Yanta were conveniently positioned immediately south of the prospect.

During their 1984 explorations, CRA damaged some local gardens. The question of who ought to receive compensation spawned a series of litigations between Yanta, Hengambu and occupants of other nearby villages.[18] While the litigation started as a compensation dispute, after a series of decisions, on appeal Judge Geoffrey Charles Lapthorne awarded the Hengambu and Yanta clans 50 per cent ownership of the 'Wafi River Prospect' area (GPNG 1985). Like *Megentse*, the peculiar circumstances of the decision shaped the resulting legal landscape. Most importantly for the considerations at hand, the case only vaguely attempted to resolve apparent discrepancies with the earlier *Megentse* cases, explaining that the

---

18 Twangala, Bupu, Omalai, Piu and Parakles.

50/50 decision 'should not be construed so as to diminish any rights or claims the Bobop[19] [sic] people may have to the land' the decision rules on (GPNG 1985). It is difficult to know whether this is the case, as the Megentse land was never clearly defined in the 1981 and 1982 decisions.

The two sets of court cases—*Megentse* and *50/50*—set a confused and arguably contradictory precedent. Babuaf was the winning party in one, but was absent in the other. Hengambu was a party in both cases, losing one and winning another. Yanta was a winning party in one case, but was absent in another. Wampar-speakers testified in one to their historical occupation of the land but was not an official party in either case. A range of other villages were the losers of the 1984 and 1985 decisions, while other Central Watut–speaking villages go largely unmentioned in both decisions.

Regardless—or perhaps precisely because of these discrepancies—the 1980s formed a critical juncture in the history of the Wafi-Golpu region, whose consequences no number of appeals could overturn. Prior to the 1980s, it was broadly open whom the state would eventually recognise as customary landowners of the area. However, by the end of the 1980s, possibilities had narrowed significantly. Not only because of who won, but *how* the sociality of the region happened to be cut up. Some 20 years later, the 1980s are relitigated both literally and rhetorically as *the* turning point in which the subsequent legal positioning of claimants were solidified. By the end of the 1980s, the Wafi area was gifted with three new legal categories—Babuaf, Hengambu and Yanta, each with unclear membership criteria—that had become the legally recognised owners of land related to the Wafi prospect, albeit in various contradictory capacities. Those best able to speak to these categories entered the second phase as legal and physical incumbents.

Assembling Organisations

The 1980s created new legal categories, customary landowners and, with the increasing prospect of future mining, made those categories economically valuable and politically potent. They also changed the costs of competition. Having made their decisions, challenging the winners of the 1980s via Local Land Courts was no longer viable. While the *50/50*

---

19   Likely meaning Babuaf.

decision might have been loose with the rules of precedent, a repeat of such an event was unlikely to occur again. Instead, groups with real or imagined claims to Wafi-Golpu land that were excluded from the earlier decisions would need to organise in more systematic and innovative ways. Likewise, incumbents would need to organise to meet these threats.

To this end, the 2000s saw two main skirmishes between the different Wafi claimants, both characterised by unconventional, yet legalistic, approaches (Table 3.2). The first was an attempt by individuals from Piu, one of the smaller claimants on the losing end of the 1980s cases, to surreptitiously claim a SABL over the entire Wafi-Golpu area. The second, the SLTC over Wafi-Golpu, fundamentally changed the politics and alliances of Wafi-Golpu region.

Table 3.2 Legal events pertaining to the Piu SABL.

Date	Event
26 July 2001	Delegate of the Minister of Land grants Piu Land Group Inc. a SABL of the whole Wafi area.
2003	National Court reinstates Piu Land Group Inc. SABL over Wafi area.
2005	Supreme Court strikes down Piu Land Group Inc. SABL.

Source: GPNG (2005).

The following two sections recount these two periods, tracking the increasing solidification of a range of factions through repeated preparation for, and attendance of, court. Compared to the short-lived action sets of the 1980s, where prominent locals temporarily worked together to testify in court cases, the factions of the following two sections become more deliberately organised social collectives, specifically and explicitly organised for the collective action of litigating over the future gains of Wafi-Golpu. Critically, the increasing financial and logistic needs of litigation fed the need for broad coalition building and pushed power into the hands of leaders of local factions so that they could make court filings, fly to Moresby, incorporate companies and do the everyday work of litigation.

1997–2005: Land Leases and New Leaders

On 26 July 2001, a delegate of the Minister of Land surreptitiously granted Piu Land Group Inc., a corporation claiming to represent the Piu people, a SABL of the whole Wafi area.[20] A SABL is a legal mechanism for

---

20   See GPNG (2005) for a summary of events leading up to the Piu SABL.

a 'lease–lease-back' scheme, realised through the *Land Act 1996*. Legally, the Minister of Lands leases customary land from its customary owners, in order to in turn lease the land to 'a) to a person or persons; or b) to a land group, business group or other incorporated body, to whom the customary landowners have agreed that such a lease should be granted'[21] (see Filer 2011 for summary of the role of SABLs in the politics of land in PNG; and Schwoerer, this volume).

In the eyes of the beneficiaries of the 1980s cases, this was a blatant challenge to their ownership of the area, especially considering Piu was one of the parties explicitly listed on the losing side of the 1985 *50/50* decision. A flurry of complaints led to the Lands Department revoking the licence—only to have it reinstated by the National Court without the presence of any of the complainants (GPNG 2003). Appeals then pushed the case up to the Supreme Court (GPNG 2005). In order to get a clearer sense of how individuals came to lead the factions that worked to challenge the SABL, I focus on two key leaders, Thomas Nen and Bill Itamar, and how they worked together in a Babuaf–Wampar alliance to overturn the SABL.

Bill Itamar represented Wampar interests in the alliance. In 2016/17 when I met him, Bill often wore crisp, collared shirts and a driving cap typical of Morobe. He was a frequent sight on the pothole-filled Wau–Bulolo highway riding shotgun in his bright yellow PMV (public motor vehicle) going to or from the city of Lae. Bill stabilised and began formalising the political alliance between Babuaf and Wampar in the 1990s after initial uncertainty in the wake of the *50/50* decisions. According to a witness from the time,[22] when the *50/50* decision was handed down, there was disagreement between those involved in the *Megentse* cases over the appropriate course of action. Some Wampar wanted to appeal the *50/50* decision, but those in Babuaf were less certain. One of the original Wampar witnesses attempted to appeal the case himself but was rebuffed because the *Megentse* cases were awarded to Babuaf. Consequently, talk emerged among Wampar leaders about breaking off the alliance entirely.

---

21 *Land Act 1996*, Section 102.
22 Yaeng Ngawai, from conversation in Mare, 2 October 2017. All direct quotes from informants in this chapter were spoken in Tok Pisin, and translated by the author. At the time of the original *Megentse* court decisions, Ngawai was undertaking mission work in Bwana, but after returning to Mare, he began helping Go Nowa, one of the original Wampar witnesses for the *Megentse* case, with the fallout of the 1980s decisions.

As one witness complained 'they [Babuaf] got the credit but did nothing. The court would recognise that and make them number two'.[23] These plans came to a halt when Bill Itamar intervened.

Neither of Bill's parents were early prominent political figures, and nor were they particularly well-off. After his mother died when he was a child, Bill was raised by his mother's sister, who paid for Bill's school fees using money from the then-booming betelnut trade.[24] Bill did well in school, so Lae Technical College sponsored him to study clerical and business studies between 1972 and 1973. He spent some time going to and from Enga province and the capital of Port Moresby, working and studying, in addition to marrying a woman from Mare in 1978. After graduating in 1980, Bill travelled to the Eastern Highlands, beginning work as a business development officer for the provincial government before reaching high levels of public administration as provincial financial adviser in 1989 and provincial planner in the early 1990s. While working for the government in Goroka, he had a life-changing encounter with the Pentecostal Christian Life Church (CLC).

Bill recounts his life before CLC as full of indiscretion.[25] He chewed betelnut, drank all the time and frequented clubs while in Port Moresby. He also played basketball semi-professionally and was selected to go to the Pacific Games. However, during his stay in the Eastern Highlands, Bill met Pastor John Kemp who converted Bill and his wife to born-again Christianity. So, while Bill studied public finances and accounting in Goroka, he also deepened his faith.

Bill periodically returned to Mare throughout the 1980s and early 1990s, a changed man. At the time, the Lutheran Church had a spiritual monopoly in the region, and when Bill started a CLC church, the Lutheran orthodoxy resisted. According to witnesses from both sides of the schism, Lutheran followers arrived at Bill's house brandishing machetes, spears, axes and burning coconuts, forcing Bill to move his house to the very edge of the village. Nevertheless, over many years, Lutherans gradually came to accept CLC.

---

23 Interview in Mare, 2 October 2017.
24 Wampar grew betelnut as a major source of income until 2007, when an unknown pest wiped out their crop.
25 Interview with Bill Itamar, 4 October 2017, Mare.

Bill's rising religious fortunes coincided with his political ascendance. Although his 1992 run for the Huon Gulf electorate was unsuccessful, by 1997 and again in 2002 he was elected councillor of Mare. In 1997, at the peak of Bill's political rise, Mare voted on who ought to continue the legacy of the *Megentse* decision, and who ought to lead Wampar on Wafi-Golpu-related matters. Unlike others, Bill thought that Wampar should still work with Babuaf, and the assembled village voted for Bill to lead.

Bill initially worked with Peter Ngawas, another Watut man, under the name 'Babwaf'. Then around 2001, Babuaf elected Bill's future rival, Thomas Nen. In 2016, Thomas sported a goatee and frequently wore fedoras when going to and from court, having represented Babuaf since his election. Thomas comes from Dzemep, a southern Watut village, not Babuaf itself—a point of occasional contention given that he represents himself as a 'landowner' from the 'Babwaf tribe' in legal documents. Like Bill, he is highly educated, having studied development economics in the United Kingdom, as well as 'regional studies' in China during the 1980s.[26]

After returning to PNG with a Chinese wife around 1989, Thomas rose to become managing director of the PNG Forestry Authority in 1998 (GPNG 2002). While at the Forestry Authority, he became embroiled in a series of logging scandals in Western Province. According to Brain Brunton, a lawyer-consultant for Greenpeace, during his time as director Thomas travelled back and forth to China connecting Chinese timber companies interested in PNG hardwoods (1998a, 1998b). Subsequent investigations resulted in an Ombudsman Commission report that concluded that Thomas had acted incorrectly and that 'the future public re-employment of Thomas Nen must be carefully and critically reviewed' (GPNG 2002: 6). Thomas was subsequently removed from his position at the Forestry Authority in March 2002 (Canberra Friends of PNG 2002). Barely breaking his stride, three months later in June 2002 Thomas ran for election as MP of Huon Gulf, coming only 448 votes short of the lead candidate (Development Policy Centre 2020).

This unlikely pair—Thomas with his Chinese and Moresby connections and Bill with his spiritual coalition and provincial government experience—worked together against Piu's SABL. For the next few years, while Thomas filed affidavits at the Supreme Court, Bill solicited Judge Steven Awagasi, the original magistrate for the *Megentse* case, to support

---

26 Interview with Thomas Nen, 29 November 2016, Lae.

their cause (Awagasi 2004). Together, prominent men from both Mare and Babuaf signed a letter petitioning the Minister of Lands to withdraw the SABL (Nen 2004). The struggle was not inexpensive—Bill and his broader alliance gave 5,000 kina for the Supreme Court fees (Itamar 2009a). Finally, on 29 August 2005, their labours were rewarded, and the Supreme Court revoked the SABL licence (GPNG 2005).

Formal Registration

At the end of the first half of the 2000s, distinct factions had formed around the Wafi-Golpu area, forged through repeated, antagonistic interactions in litigation. Over the course of a final round of litigation over the 'SLTC', created to resolve customary disputes over Wafi-Golpu, these groupings calcified and, by the end, each faction was represented by distinct legal entities.[27] By the time I arrived in the field in 2016, all of the claimants from each of the linguistic populations touched on in this chapter had one or more registered landowner associations (LOAs), each with complementary landowner businesses, all headed by members like Bill and Thomas. Each of these LOAs operated under a similar logic: each had an outspoken chairman,[28] who undertook the majority of the visible labour of political competition, an associated group of 'directors' more active in the association, supported by a broad range of 'members' who occasionally attended large celebrations and feasts, voted and paid membership fees.

The pervasiveness of LOAs in the Wafi-Golpu area is a function of two features. First, they are the result of state and developer demands for clear representatives of landowners, in a suitable social form, to negotiate with. As the head of the mine owner's community affairs department explained succinctly:

> The idea is that you start early, do awareness, and narrow and narrow and narrow down how many people will be there [at the Development Forum] until there are just the leaders to speak for everyone, and the government feels it is representative enough.

---

27 Court cases were by no means the only reason for the formation of LOAs; before the 2000 cases, Hengambu and Yanta had all begun organising into LOAs in anticipation of mining benefits. The Hengambu Landowner Association was founded in 2000, while the Yanta Landowner Association was founded in approximately 1998. However, the legal conflicts over the 2000s saw these forms of association become standard for almost every claimant group in the region.

28 The chairmen of LOAs were exclusively male.

> To make sure the biggest landowners are there. If not, every Tom, Dick and Harry will show up and nothing will get done. You know how PNG is.[29]

The ease in which the company and state representatives discuss 'representative landowners' belies the reality that LOAs, or the various other legal entities that operate around mining projects in PNG, are not reflections of local social affiliations, but are specific responses to the demands from the company and the state for more tractable social units to negotiate with. As Jorgensen stresses, the creation of such formalised, corporate entities are:

> an exercise in the creation of legal fictions fulfilling the state's need to delineate landowners for the purposes of concluding mining agreements, and a solution hinges upon formulated identities in a way that satisfies the state's interests in legibility by *making clans that the state can 'find'*. (Jorgensen 2007: 66, emphasis in original)

These formalised entities also have the dual function of acting as having emerged from the legal back-and-forth recounted so far, as the primary units of factional competition around Wafi-Golpu.

Table 3.3 Legal events pertaining to the Wafi-Golpu SLTC.

Date	Event
24 September 2008	Minister of Justice founds the SLTC over customary ownership of Wafi Prospect Land.
19 January 2011	Acting Governor-General revokes the SLTC commissioners, disbanding the commission.
6 November 2011	National Court rules disbanding of the SLTC a breach of natural justice, and demands reinstatement of the commission.
October 2018	Supreme Court strikes down National Court ruling, upholding the disbanding of SLTC.

Source: GPNG (2011, 2018).

2008–2018: Special Land Titles Commission

PNG's legacy of colonial land laws, coupled with the pervasiveness of customary ownership, has left the country with a unique land dispute process. Prior to 1975, the Land Titles Commission (LTC) held exclusive jurisdiction over all customary land disputes. Since the passage of the

---

29   Interview with David Masani, 24 November 2016, Gabsongkeg.

Land Disputes Act 1975, a separate hierarchy of courts—local land courts, district land courts and provincial land courts—have adjudicated land matters,[30] leaving the LTC as a vestigial quasi-judicial tribunal that acts as a special arbitrator when the head of state explicitly invokes the commission.[31] Doing so transfers the jurisdiction of a disputed area of land from the lands courts to the LTC. Such a transfer occurred on 24 September 2008 in response to pressure from parties, including Bill and Thomas, that were dissatisfied with the *50/50* decision. The result was the formation of the SLTC to resolve the customary land disputes over Wafi-Golpu once and for all.[32]

The SLTC began at a particularly tense moment for the Babuaf–Wampar alliance. For a brief moment following the disbanding of the Piu SABL, it seemed that the union would hold. In 2005, at a large meeting attended by people from the villages of Babuaf and Mare, the two groups agreed that the Kutut Development Corporation and the Saab Development Corporation would work together under the name 'Babwaf' (Itamar et al. 2005). However, differences between Thomas and Bill began to break the Wampar and Babuaf alliance apart, with Thomas pushing to separate from Wampar, arguing that Bill, and Wampar more broadly, had no claim to the earlier *Megentse* decisions. A year after the unifying meeting, Thomas and Bill agreed to split the organisation in two, with Thomas leading the Wale Babwaf Landowner Association focusing on the Watut, and Bill leading Wampar with the Babwaf Saab Landowner Association (Itamar 2009b).

In 2007, both Bill and Thomas unsuccessfully ran for election as MP for Huon Gulf. With the appointment of the SLTC, these divisions broke into a full-scale legal conflict. The SLTC reshaped the politics of the whole Wafi area. Whatever temporary unity Bill's alliance had managed to contain within the Wampar region broke. Among Wampar-speakers, six different parties registered, five from the village of Mare. The other claimants from the area around Wafi were no less divided and the SLTC had 31 different claimants in total, each alleging exclusive ownership of the Wafi-Golpu project area.

---

30   In theory, this means that PNG has a separate hierarchy of courts for dealing with land matters, distinct from the conventional Commonwealth court hierarchy. In practice, the judges that serve in the land courts are frequently the same as those that serve in the conventional ones.

31   *Land Disputes Settlements Act 1975*, Section 4.

32   See GPNG (2011) for a summary of events that led to the dismissal of the SLTC.

As beneficiaries of the *50/50* decision, LOAs representing Hengambu and Yanta were consistently hostile to the SLTC as a potential threat to their position. Since its inception, Hengambu and Yanta leaders threatened to physically shut down the Wafi-Golpu prospect if the government did not end the SLTC. By mid-April 2010, as the SLTC prepared to demarcate boundaries between landowners, these threats rose in intensity to the point where the Minister for Mines flew by helicopter to the Wafi exploration camp to meet with the complainants. A little under a year later, the Acting Governor-General revoked the SLTC commissioners and disbanded the commission. The excluded parties' hope for inclusion into the project had a brief opening when the National Court ruled the ending of the SLTC was a breach of natural justice, which the state promptly appealed to the Supreme Court (GPNC 2011).

Since the SLTC was disbanded, Bill and his allies have been consumed by their quest to reinstate the commission and, more broadly, for the government and the mine owners to recognise their claim as customary landowners. In June 2018, in a coalition of the ousted and the ignored, Bill, former leaders from the Yanta and Hengambu LOAs, representatives from the wider Watut area, allied members of other Wampar villages and representatives from villages along the proposed Wafi-Golpu slurry pipeline joined forces and formed the Wafi-Golpu Landowner Mine Association and registered Wafi-Golpu Holdings Limited. This umbrella landowner association and landowner company claimed to represent all the landowners of Wafi-Golpu, a 'new association that covers everybody, from the pit down to the wharf' (EMTV Online 2018).

However, on 11 July 2018, the Development Forum began with no invitation forthcoming for this umbrella association. Any remaining hopes were dashed in October 2018 when the Supreme Court handed down its decision, quashing the earlier reinstatement of the SLTC, arguing that the commission had no grounds for ruling on already decided cases from the 1980s (GPNG 2018). As such, at the time of writing, Wafi-Golpu sits in an ambiguous legal status quo. Now the Mineral Resource Authority, working with the mine owners, are attempting to convince the three winning claimants (Babuaf, Hengambu and Yanta) to form one unified LOA, with an associated landowner company, to sign the memorandum of agreement at the Development Forum. Whoever heads the resulting LOA will wield significant power over how MRBs are distributed locally.

Conclusion

> Now, here, you see, it takes all the running you can do, to keep in
> the same place.
> —the Red Queen to Alice, in Lewis Carroll's *Through the Looking-
> Glass* (Carroll 1871)

The legal history recounted here may seem Sisyphean. With the dust
settled and Wafi-Golpu entering the final steps of the licensing process,
the legal result at the time of writing in 2022 is the same as that of the
1980s. The state has still not mapped the boundaries of the Megentse
region, the contradictory precedent between the 1980s cases remains
unresolved and the mine developer and state agencies continue to split the
difference of the 1980s court cases and work equally with representatives
of Babuaf, Hengambu and Yanta. Scholars familiar with the patterns
of resource brokerage in PNG will likely find the figures of Bill Itamar
and Thomas Nen familiar; the clientelistic relations they (hope) to foster
mirror the forms of brokerage discussed by Monica Minnegal and Peter
D. Dwyer (Chapter 4, this volume).

However, an impression of stasis misses the social consequences of the
last 40 years of struggle—the work required to 'keep in the same place',
as the Red Queen said to Alice. In the 1980s, court participants were
temporary action sets gathered together to provide specific testimony in
court. For these early cases, knowing Tok Pisin was sufficient to confer
significant advantages. However, merely to *maintain* the results of the
early 1980s, significant work had to be undertaken: fending off SABLs,
litigation in Moresby, assembling pan-village coalitions, and formal
registration of LOAs. By the end of the 2010s, the Wafi-Golpu area
became characterised by stratified factions, topped by well-connected and
educated older men, linked to their followers in networks of promised
clientelism, specifically designed for factional competition. These factions
are connected to a range of legal entities, including LOAs and landowner
companies, perfectly set up for lopsided distribution of MRBs.

By tracing this process, I argued that legal competition over MRBs
constitutes a form of stratifying factional competition due to three key
features:

1. Legal action has substantial social, economic and political requirements, limiting the range of individuals who have the capacity to draw together the necessary coalitions to participate in such competitions.

2. These requirements progressively escalate while driving positive feedback loops that solidify existing advantages.

3. Legal decisions and state expectations drive a dynamic process whereby factions increasingly represent themselves in the form of legalised associations whose very names and affiliations are shaped by earlier events.

Owing to these features, I argued that legal competition for MRBs drives organisational stratification among involved factions.[33]

Government and developer representatives frequently blame LOAs and their leaders for the adverse distributional outcomes of mining benefits. In *The National,* one of PNG's national newspapers, Sean Ngansia, the executive manager of the Development Coordination Division in the Mineral Resources Authority, complained:

> We don't necessarily manage royalties on landowners' behalf ... [royalties] is usually given directly to the landowners through their landowner associations. The issue now is really about how these monies are managed. You will find that in Hidden Valley and all the other mines, the landowner association leaders are not managing their royalties well. There's a lot of misuse and mismanagement. These leaders also do not report to their people and that's where the problem is. (Ngansia 2018)

There is indeed woefully inadequate transparency about mining benefits distribution, a point that observers of the PNG extraction industries have repeatedly made (Sagir 2001; Koyama 2004; Haley and May 2007; Filer 2012). However, these debates can involve a sleight-of-hand, signalled by the above quote's complaints of 'misuse and mismanagement'. Developer and government complaints about a lack of coordination and leadership

---

33   My emphasis on increasing political rigidity should not be read as immunity to change. Bill Itamar displaced earlier Wampar leaders in the 1990s, and representatives from Piu managed to secure a SABL in 2001 despite their disadvantageous position. However, such flexibility was limited, and becomes increasingly so as time goes on. Critically, the ability to undertake such upsets (or defend from them) was not equally distributed, favouring male leaders with educational, financial and social resources. Pulling together a wide range of interests favoured precisely those individuals with the resources to make credible commitments to their followers and peers.

among impacted communities misinterpret what that coordination and leadership are *for*. Rather than seeing contested landowner representation as a case of failed coordination, it is more accurate to see factions as the result of *successful* coordination for a specific, winner-takes-almost-all competition not of the participants' own making.

References

Arellano-Yanguas, J., 2011. 'Aggravating the Resource Curse: Decentralisation, Mining and Conflict in Peru.' *Journal of Development Studies* 47(4): 617–638. doi.org/10.1080/00220381003706478

———, 2017. 'Inequalities in Mining and Oil Regions of Andean Countries.' *Journal of Development Studies* 6(2): 98–122. doi.org/10.26754/ojs_ried/ijds.255

Arthur, W.B., 1994. *Increasing Returns and Path Dependence in the Economy.* Ann Arbor: University of Michigan Press. doi.org/10.3998/mpub.10029

Auty, R.M., 1993. *Sustaining Development in Mineral Economies: The Resource Curse Thesis.* London; New York: Routledge.

Awagasi, S., 2004. 'Babwaf v. Engabu. Awagasi to Itamar. 1st March 2004.' Signed Letter, Annex C to the affidavit of Bill Itamar in the Land Title Commission, Application No. 2008/30 (1-29) 2004.

Badeeb, R.A., L. Hooi and J. Clark, 2017. 'The Evolution of the Natural Resource Curse Thesis: A Critical Literature Survey.' *Resources Policy* 51(March): 123–134. doi.org/10.1016/j.resourpol.2016.10.015

Bailey, F.G., 1969. *Stratagems and Spoils: A Social Anthropology of Politics.* Oxford: Blackwell.

Bainton, N., 2009. 'Keeping the Network out of View: Mining, Distinctions and Exclusion in Melanesia.' *Oceania* 79(1): 18–33. doi.org/10.1002/j.1834-4461.2009.tb00048.x

———, 2010. *The Lihir Destiny: Cultural Responses to Mining in Melanesia.* Canberra: ANU E Press (Asia-Pacific Environment Monographs). doi.org/10.22459/LD.10.2010

Bainton, N. and M. Macintyre, 2013. '"My Land, My Work": Business Development and Large-Scale Mining in Papua New Guinea.' In F. McCormack and K. Barclay (eds), *Engaging with Capitalism: Cases from Oceania*. Bingley: Emerald Group Publishing Limited (Research in Economic Anthropology 33). doi.org/10.1108/S0190-1281(2013)0000033008

Bainton, N. and J.R. Owen, 2019. 'Zones of Entanglement: Researching Mining Arenas in Melanesia and Beyond.' *The Extractive Industries and Society* 6(3): 767–774. doi.org/10.1016/j.exis.2018.08.012

Ballard, C., 1993a. 'Golpu (Wafi) Prospect Social Mapping Study.' Report prepared for CRA Minerals (PNG) Pty Ltd, Port Moresby. Port Moresby: Unisearch PNG Pty Ltd.

———, 1993b. 'Babwaf and Piu: A Background Study.' Report Prepared for CRA Minerals (PNG) Pty Ltd, Port Moresby. Port Moresby: Unisearch PNG Pty Ltd.

Ballard, C. and G. Banks, 2003. 'Resource Wars: The Anthropology of Mining.' *Annual Review of Anthropology* 32: 287–313. doi.org/10.1146/annurev.anthro.32.061002.093116

Banks, G., 2005. 'Linking Resources and Conflict the Melanesian Way.' *Pacific Economic Bulletin* 20(1): 8.

———, 2019. 'Extractive Industries in Melanesia.' In E. Hirsch and W. Rollason (eds), *The Melanesian World* (1st edition). London; New York: Routledge. doi.org/10.4324/9781315529691-30

Barnes, J.A., 1969. 'Networks and Political Process.' In M.J. Swartz (ed.), *Local-Level Politics: Social and Cultural Politics*. London: University of London Press.

Beer, B., 2006. 'Stonhet and Yelotop: Body Images, Physical Markers and Definitions of Ethnic Boundaries in Papua New Guinea.' *Anthropological Forum* 16(2): 105–122. doi.org/10.1080/00664670600768284

Boissevain, J., 1978. *Friends of Friends: Networks, Manipulators and Coalitions* (Reprint). Oxford: Blackwell (Pavilion Series).

Brumfiel, E., 1992. 'Distinguished Lecture in Archeology: Breaking and Entering the Ecosystem—Gender, Class, and Faction Steal the Show.' *American Anthropologist* 94(3): 551–567. doi.org/10.1525/aa.1992.94.3.02a00020

———, 1994. 'Factional Competition and Political Development in the New World: An Introduction.' In E. Brumfiel and J. Fox (eds), *Factional Competition and Political Development in the New World*. Cambridge: Cambridge University Press. doi.org/10.1017/CBO9780511598401

Brumfiel, E. and J. Fox (eds), 1994. *Factional Competition and Political Development in the New World*. Cambridge: Cambridge University Press. doi.org/10.1017/CBO9780511598401

Brunton, B., 1998a. 'Forest Update.' Report for Greenpeace Pacific, 1998.

———, 1998b. 'Papua New Guinea Forest Update.' Report for *Pacific Islands Report*, 1998.

Bujra, J., 1973. 'The Dynamics of Political Action: A New Look at Factionalism.' *American Anthropologist* 75(1): 132–152. doi.org/10.1525/aa.1973.75.1.02a00080

Canberra Friends of PNG, 2002. *Partners in Crime: The Political Web That Supports the Illegal Kiunga Aiambak Timber Project*. Viewed 10 January 2022 at: pngiportal.org/directory/cfpa2002-pdf

Capoccia, G. and R. Kelemen, 2007. 'The Study of Critical Junctures: Theory, Narrative, and Counterfactuals in Historical Institutionalism.' *World Politics* 59(3): 341–369. doi.org/10.1017/S0043887100020852

Carroll, L., 1871. *Through the Looking Glass, and What Alice Found There*. London: Macmillan.

Chacon, R. and R. Mendoza (eds), 2016. *Feast, Famine, or Fighting? Multiple Pathways to Social Complexity*. New York: Springer. doi.org/10.1007/978-3-319-48402-0

Church, W., 2019. 'Changing Authority and Historical Contingency: An Analysis of Socio-Political Change in the Colonial History of the Markham Valley (Papua New Guinea).' *Paideuma* 65: 61–86.

Cox, J., 2018. *Fast Money Schemes: Hope and Deception in Papua New Guinea*. Bloomington: Indiana University Press. doi.org/10.2307/j.ctv6mtfjm

Development Policy Centre, 2020. *PNG Elections Database*. Canberra: The Australian National University. Viewed 26 January 2022 at: devpolicy.org/pngelections/

Dwyer, P. and M. Minnegal, 1998. 'Waiting for Company: Ethos and Environment Among Kubo of Papua New Guinea.' *The Journal of the Royal Anthropological Institute* 4(1): 23–42. doi.org/10.2307/3034426

EMTV Online, 2018. 'Landowners Unite to Form Wafi-Golpu Landowner Mine Association.' YouTube. Viewed 26 January 2022 at: www.youtube.com/watch?v=eidtdI-qvEY

Filer, C., 1990. 'The Bougainville Rebellion, the Mining Industry and the Process of Social Disintegration in Papua New Guinea.' *Canberra Anthropology* 13(1): 1–39. doi.org/10.1080/03149099009508487

———, 1992. 'The Escalation of Disintegration and the Reinvention of Authority.' In M. Spriggs and D. Denoon (eds), *The Bougainville Crisis: 1991 Update*. Canberra: The Australian National University, Department of Social and Political Change.

———, 1997. 'Compensation, Rent and Power in Papua New Guinea.' In S. Toft (ed.), *Compensation for Resource Development in Papua New Guinea*. Port Moresby; Canberra: PNG Law Reform Commission and The Australian National University (Pacific Policy Papers 24).

———, 2011. 'The Political Construction of a Land Grab in Papua New Guinea.' Canberra: The Australian National University, Crawford School of Economics and Government (READ Pacific Discussion Paper 1).

———, 2012. 'The Development Forum in Papua New Guinea: Evaluating Outcomes for Local Communities.' In M. Langton and J. Longbottom (eds), *Community Futures, Legal Architecture: Foundations for Indigenous Peoples in the Global Mining Boom*. Oxford: Routledge.

Firth, R., 1957. 'Introduction to Factions in Indian and Overseas Indian Societies.' *British Journal of Sociology* 8: 291–295.

Fischer, H., 1963. *Watut: Notizen zur Kultur eines Melanesier-Stammes in Nordost Neuguinea* [Watut: Notes on the Culture of a Melanesian Tribe in Northeast New Guinea]. Kulturgeschichtliche Forschungen Bd. 10. Braunschweig: Albert Limbach Verlag.

———, 1992. *Weisse und Wilde: Erste Kontakte und Anfänge der Mission* [White Men and Wild People: First Contacts and the Beginnings of Evangelisation]. Berlin: Reimer (Materialien zur Kultur der Wampar, Papua New Guinea 1).

———, 2013. 'Woher wir kamen: Moderne Elemente zur Herkunftsgeschichte der Wampar, Papua-Neuguinea [Where We Came From: Modern Elements on the History of the Origin of the Wampar].' *Sociologus* 63(1–2): 125–45.

Fischer, H. and B. Beer, 2021. *Wampar–English Dictionary*. Canberra: ANU Press. doi.org/10.22459/WED.2021

Galanter, M., 1974. 'Why the "Haves" Come out Ahead: Speculations on the Limits of Legal Change.' *Law & Society Review* 9(1): 95–160. doi.org/10.2307/3053023

Gilberthorpe, E., 2007. 'Fasu Solidarity: A Case Study of Kin Networks, Land Tenure, and Oil Extraction in Kutubu, Papua New Guinea'. *American Anthropologist* 109(1): 101–112. doi.org/10.1525/aa.2007.109.1.101

Gilberthorpe, E. and E. Papyrakis, 2015. 'The Extractive Industries and Development: The Resource Curse at the Micro, Meso and Macro Levels.' *The Extractive Industries and Society* 2(2): 381–390. doi.org/10.1016/j.exis. 2015.02.008

Goldman, L., 2007. 'Incorporating Huli: Lessons from the Hides Licence Area.' In J. Weiner and K. Glaskin (eds), *Customary Land Tenure and Registration in Australia and Papua New Guinea: Anthropological Perspectives*. Canberra: ANU E Press (Asia-Pacific Environment Monographs). doi.org/10.22459/ CLTRAPNG.06.2007.06

Golub, A., 2014. *Leviathans at the Gold Mine: Creating Indigenous and Corporate Actors in Papua New Guinea*. Durham: Duke University Press. doi.org/10.1515/ 9780822377399

GPNG (Government of Papua New Guinea), 1981. *Babwaf v Engabu*. Lae Local Land Court Record of Proceedings, Unknown No. of 1981.

———, 1982. *Engabu v Babwaf*. Morobe Province District Court Record of Proceedings, Unknown No. of 1982.

———, 1985. *Yanta Clan v Hengabu Clan*. Lae District Land Court Record of Proceedings, No. 2 of 1985.

———, 1994. *Yanta v Engambu, Twangala, Bupu, Omalai, Piu*. Mumeng Local Land Court Record of Proceedings, No. 1 of 1984.

———, 2002. *Investigation into a Decision of the National Forest Board to Award Kamula Doso to Wawoi Guavi Timber Company (a Subsidiary of Rimbunan Hijau) as an Extension of the Wawoi Guavi Timber Resource Permit*. Port Moresby: Ombudsman Commission of Papua New Guinea.

———, 2003. *Piu Land Group Inc. v Sir Michael Somare and Others*. Unreported National Court judgment, OS No. 662 of 2003.

———, 2005. *Yanta Development Association Inc v Piu Land Group Inc*. Papua New Guinea Supreme Court Record of Proceedings, SC798, No. 24 of 2005.

———, 2011. *Nen and Others v Somare and Others*. National Court of Papua New Guinea Summary of Judgment, OS(JR) No. 156 of 2011.

———, 2018. *Somare v Nen*. Supreme Court of Papua New Guinea Record of Proceedings, SC1722, No. 81 of 2018.

Griffin, J., 1990. 'Bougainville is a Special Case.' In R. May and M. Spriggs (eds), *The Bougainville Crisis*. Bathurst: Crawford House Press.

Gulliver, P., 1971. *Neighbours and Networks: The Idiom of Kinship in Social Action among the Ndendeuli of Tanzania*. Berkeley: University of California Press. doi.org/10.1525/9780520317574

Haley, N. and R. May (eds), 2007. *Conflict and Resource Development in the Southern Highlands of Papua New Guinea*. Canberra: ANU E Press. doi.org/10.22459/CRD.11.2007

Hayden, B., 1995. 'Pathways to Power: Principles for Creating Socioeconomic Inequalities.' In T.D. Price and G.M. Feinman (eds), *Foundation of Social Inequality*. New York: Plenum Press (Fundamental Issues in Archaeology). doi.org/10.1007/978-1-4899-1289-3_2

Itamar, B., 2009a. 'Annexure F.' Affidavit of Bill Itamar to Special Land Titles Commission over Wafi-Golpu. Huon Gulf District.

———, 2009b. Affidavit of Bill Itamar in the Land Titles Commission. Huon Gulf District.

Itamar, B., P. Ngamus and J. Marvin Jackson, 2005. 'Saab Kundut (Babwaf) Executive Board Meeting Minutes. 28 May 2005.'

Jacka, J.K., 2015. 'Uneven Development in the Papua New Guinea Highlands: Mining, Corporate Social Responsibility, and the "Life Market".' *Focaal* 73: 57–69. doi.org/10.3167/fcl.2015.730105

Jebens, H. (ed.), 2004. *Cargo, Cult and Culture Critique*. Honolulu: University of Hawai'i Press. doi.org/10.1515/9780824840440

Jorgensen, D., 1997. 'Who and What Is a Landowner? Mythology and Marking the Ground in a Papua New Guinea Mining Project.' *Anthropological Forum* 7(4): 599–627. doi.org/10.1080/00664677.1997.9967476

———, 2007. 'Clan-Finding, Clan-Making and the Politics of Identity in a Papua New Guinea Mining Project.' In J.F. Weiner and K. Glaskin (eds), *Customary Land Tenure and Registration in Australia and Papua New Guinea: Anthropological Perspectives*. Canberra: ANU E Press (Asia-Pacific Environment Monographs). doi.org/10.22459/CLTRAPNG.06.2007.04

Kapferer, B. (ed.), 1976. *Transaction and Meaning: Directions in the Anthropology of Exchange and Symbolic Behavior*. Philadelphia: Institute for the Study of Human Issues (ASA Essays in Social Anthropology 1).

Kirsch, S., 2018. *Engaged Anthropology: Politics Beyond the Text*. Berkeley: University of California Press. doi.org/10.1525/california/9780520297944.001.0001

Koyama, S.K., 2004. 'Reducing Agency Problems in Incorporated Land Groups.' *Pacific Economic Bulletin* 19(1): 20–32.

———, 2005. '"Black Gold or Excrement of the Devil"? The Externalities of Oil Production in Papua New Guinea.' *Pacific Economic Bulletin* 20(1): 14–26.

Lawrence, P., 1964. *Road Belong Cargo: A Study of the Cargo Movement in the Southern Madang District New Guinea*. Prospect Heights: Waveland Press.

Lindstrom, L., 1993. *Cargo Cult: Strange Stories of Desire from Melanesia and Beyond*. Honolulu: University of Hawai'i Press.

Macintyre, M., 2003. 'Petztorme Women: Responding to Change in Lihir, Papua New Guinea.' *Oceania* 74(1–2): 120–134. doi.org/10.1002/j.1834-4461.2003.tb02839.x

Minnegal, M. and P. Dwyer, 2017. *Navigating the Future: An Ethnography of Change in Papua New Guinea*. Canberra: ANU Press (Asia-Pacific Environment Monographs). doi.org/10.22459/NTF.06.2017

MMSD (Mining Minerals and Sustainable Development Project), 2002. *Breaking New Ground: Mining, Minerals and Sustainable Development*. London: International Institute for Environment and Development.

Nen, T., 2004. 'Affidavit in Support.' Application for Leave to Apply for Judicial Review of Piu Land Group Inc. v Sir Michael Somare & Ors, 2003.

Ngansia, S., 2018. 'Basil Wants Mine Not to Repeat Past Mistakes.' *The National*, 13 July 2018.

Nicholas, R., 1965. 'Factions: A Comparative Analysis.' In M. Banton (ed.), *Political Systems and the Distribution of Power*. London: Tavistock Publications Limited (ASA Monographs 1).

Pierson, P., 2000. 'Increasing Returns, Path Dependence, and the Study of Politics.' *The American Political Science Review* 94(2): 251–267. doi.org/10.2307/2586011

Roscoe, P., 1993. 'Practice and Political Centralisation: A New Approach to Political Evolution [and Comments and Reply].' *Current Anthropology* 34(2): 111–140. doi.org/10.1086/204149

Sachs, J. and A. Warner, 1995. 'Natural Resource Abundance and Economic Growth.' Cambridge, MA: National Bureau of Economic Research (Working Paper 5398). doi.org/10.3386/w5398

Sagir, B.F., 2001. 'The Politics of Petroleum Extraction and Royalty Distribution at Lake Kutubu.' In A. Rumsey and J. Weiner (eds), *Mining and Indigenous Lifeworlds in Australia and Papua New Guinea*. Adelaide: Crawford House Press.

Schorch, P. and A. Pascht, 2017. 'Reimagining Oceania through Critical Junctures—Introduction'. *Oceania* 87(2): 114–123. doi.org/10.1002/ocea.5156

Silverman, M. and R. Salisbury (eds), 1977. *A House Divided? Anthropological Studies of Factionalism*. St. John's, Newfoundland and Labrador: Institute of Social and Economic Research, Memorial University of Newfoundland (Social and Economic Papers).

Skrzypek, E., 2020. *Revealing the Invisible Mine: Social Complexities of an Undeveloped Mining Project*. New York: Berghahn Books.

Songer, D.R. and R. Sheehan, 1992. 'Who Wins on Appeal? Upperdogs and Underdogs in the United States Courts of Appeals.' *American Journal of Political Science* 36(1): 235–258. doi.org/10.2307/2111431

Strathern, A., 1991. '"Company" in Kopiago.' In A. Pawley (ed.), *Man and a Half: Essays in Pacific Anthropology and Ethnobiology in Honour of Ralph Bulmer*. Auckland: Polynesian Society (Memoir 48).

Stürzenhofecker, G., 1994. 'Visions of a Landscape: Duna Pre-Meditations on Ecological Change.' *Canberra Anthropology* 17(2): 27–47. doi.org/10.1080/03149099409508418

Weiner, J., 2013. 'The Incorporated What Group: Ethnographic, Economic and Ideological Perspectives on Customary Land Ownership in Contemporary Papua New Guinea.' *Anthropological Forum* 23(1): 94–106. doi.org/10.1080/00664677.2012.736858

Weiner, J. and K. Glaskin (eds), 2007. *Customary Land Tenure and Registration in Australia and Papua New Guinea: Anthropological Perspectives*. Canberra: ANU E Press (Asia-Pacific Environment Monographs). doi.org/10.22459/CLTRAPNG.06.2007

Wheeler, S., B. Cartwright, R. Kagan and L. Friedman, 1987. 'Do the "Haves" Come out Ahead? Winning and Losing in State Supreme Courts, 1870–1970.' *Law & Society Review* 21(3): 403–446. doi.org/10.2307/3053377

Wiessner, P., 2002. 'The Vines of Complexity.' *Current Anthropology* 43(2): 233–269. doi.org/10.1086/338301

Worsley, P., 1957. *The Trumpet Shall Sound: A Study of 'Cargo' Cults in Melanesia*. London: Macgibbon and Kee.

4

The Broker: Inequality, Loss and the PNG LNG Project

Monica Minnegal and Peter D. Dwyer

Introduction

People of the southern highland fringes of Papua New Guinea (PNG) have always interacted with other worlds—those inhabited by spirits, or accessed through dreams—in pursuit of desired resources (Weiner 1988; Knauft 1998). But the arrival of colonisers, missionaries, prospectors and others has created awareness of previously unimagined worlds and revealed new forms of wealth to desire. More recently still, technologies such as mobile phones and associated social media platforms have introduced new modes of engagement between local, national and global worlds (Foster and Horst 2018).

In this chapter, we trace processes and consequences associated with one man's ventures into those new worlds, and the shifting motivations and mechanisms that framed his journey. Bob Resa has played a crucial role in brokering relationships between Febi and Kubo people from tributary watersheds of the upper Strickland River (Western Province) and others who, it seems, control access to the possible futures that those people now imagine for themselves.[1]

---

1   In contexts where there may be ambiguity about a person's intentions, and potential disagreement with respect to the morality and worth of actions and outcomes, Febi and Kubo people generally refrain from publicly naming a person whose behaviour may be judged in a negative light (Minnegal and Dwyer 2017: 107–9). The identity of the 'unnamed' person will be known to an 'in group',

'Brokers' are people who connect providers and consumers of knowledge and other resources, shape the flow of these through systems and, in most cases, seek to accrue personal benefit by doing so. To be effective, then, brokers must have connections with both sets of people, have some familiarity with the language, practices and value systems of both, and be able to move between the spaces where each operates. Recent interest in the role of brokers (James 2011; Lindquist 2015; Goodhand et al. 2016; Meehan and Plonski 2017; de Jong 2018) has focused on the forms brokerage takes in different kinds of geopolitical space—at borders or frontiers, in weak and strong states, colonial and postcolonial settings. This chapter, in contrast, through the lens of a single life history, focuses on the multiple modes of brokerage that may emerge in a single community, at the intersections of different socio-political spaces (Lindquist 2015: 873). By following the trajectory of a particular broker as he traverses those spaces, we reveal some of the frictions and contradictions between domains that have shaped his journey. We show, too, how his endeavours both contributed to differentiating the domains that he purported to bridge and enhanced social inequalities in his home communities. Brokers may be powerful, but they are also morally ambiguous individuals—people who cross social boundaries and whose motives and loyalties are thus always open to question (Lindquist 2015: 870; de Jong 2018; Severs and de Jong 2018). Ultimately, then, as in the case we describe, brokers may experience a personal sense of alienation, failure and loss.

We begin by setting the scene, geographically and historically, and positioning our approach within the broader literature on brokers and brokerage in and beyond PNG. We then present Bob Resa's story through the past 40 years, first tracing growth in his power and influence as he sought out domains that might hold the promise of wealth and well-being for his people, and then turning to a subsequent decline in influence as a new generation, and new modes of communication, began to redefine worlds that hold the key to desired futures. Finally, we reflect on implications for those who take on such roles, and those who look to them to deliver the hoped-for 'good life' (Robbins 2013), as movements of people—not just of ideas or resources—begin to reshape imaginings of what that life might entail.

---

but 'not naming' allows the two parties to maintain a semblance of amicable relations until the cause for concern either spills over or dissipates, while simultaneously reducing the risk to the aggrieved party of ensorcellment. Naming is, in a sense, a last resort. In this paper, we use pseudonyms as acknowledgement of local practice.

Background

Our interest in the story of Bob Resa has been stimulated, in part, by increasing tensions in recent years over the role and performance of brokers negotiating benefits from the massive PNG Liquefied Natural Gas (LNG) project. Through 2018, the PNG Department of Petroleum and Energy (DPE) organised several Landowner Beneficiary Identification (LOBID) exercises in an effort to resolve competing claims to a share of benefits by people asserting association with particular PNG LNG licence areas. But these were merely the latest in a series of such exercises—social mapping studies, landowner identification studies, clan vetting forums, alternative dispute resolution hearings—that have been undertaken in the past decade with little sign of final resolution (see Filer 2019 for an excellent overview).

As new LNG projects are envisaged in Gulf and Western provinces, and agreements for mining projects across the country are challenged (Bainton and Banks 2018), conflicts over landowner identification and over the right to speak for potential beneficiaries proliferate. Effective brokers in such contexts depend for their authority on recognition accorded by both putative landowners and those bureaucrats and others who have the power to declare beneficiary status. Aspiring brokers must convince all sides that they have the capacity to deliver the 'best possible' deal. But that claim itself may be contested, particularly where the potential exists, or is presumed to exist, for brokers themselves to accrue significant personal benefit through arbitrage.

For the Febi and Kubo-speaking people of PNG's Western Province (Figure 4.1), the people with whom Bob Resa lived much of his life, the most recent negotiations must be understood against a more general history of struggle to secure access to desired resources. Those resources were imagined to be controlled by representatives of the state, church and markets—institutions that operate at scales much larger than the kinship networks in which Febi and Kubo were, until recently, entangled as subsistence hunter-horticulturalists (Dwyer and Minnegal 1992). This struggle is most evident now in the efforts people make to render themselves visible to both the state and multinational corporations, doing so in the hope that they will be eventually officially recognised as landowners eligible for a share of the benefits that extraction of oil, gas, gold or timber may bring (Minnegal et al. 2015; Minnegal and

Dwyer 2017). But agreeing to the extraction of resources from their land is not the only means these people pursue in seeking to access the wealth and well-being they desire. Like people elsewhere in PNG, they seek recognition as citizens entitled to access government services, and as worthy souls who warrant support from the church in their efforts to lead 'good' lives (Gewertz and Errington 2016; Cox 2018). They also actively pursue opportunities to engage with both local and more distant markets as producers, not merely as owners, of desired goods and services. These different identities, and distinct domains of exchange, frame the different modes of brokerage that Bob Resa has sought to mobilise.

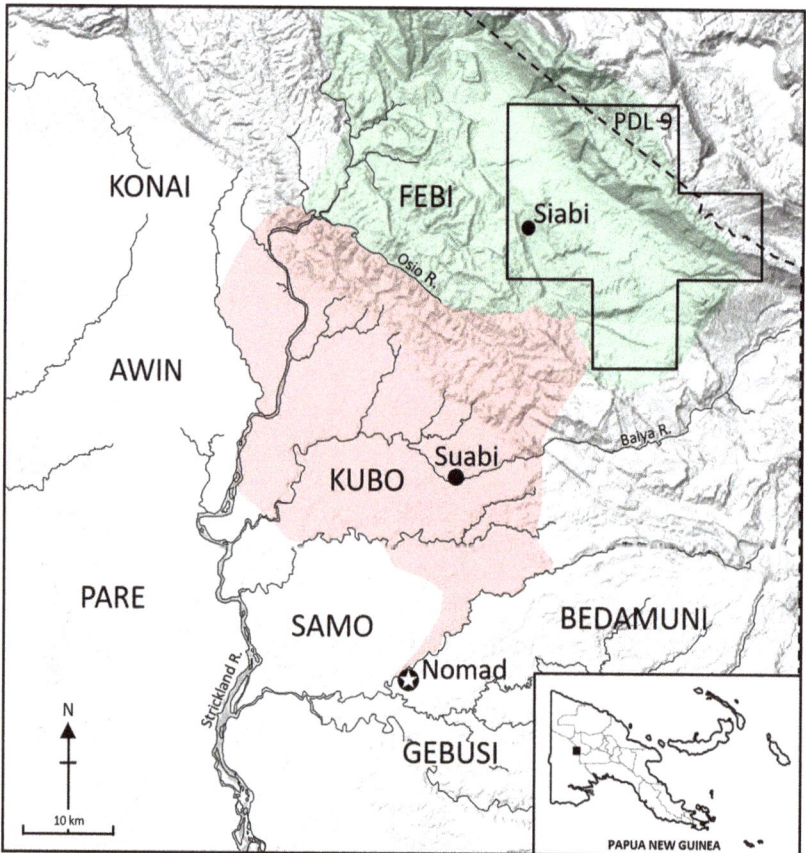

Figure 4.1 Map showing Febi and Kubo territories and location of Juha (Petroleum Development Licence area PDL 9).

Source: M. Minnegal and P.D. Dwyer.

Brokerage in New Guinea and Beyond

Like other peoples of the southern highland fringes of PNG, Febi and Kubo valorised those few men who were able to cross over into the world of spirits and, through relationships established with those met there, gain access to desired resources—particularly game—on behalf of the human communities in which they lived (Schieffelin 1976, 1977; Sørum 1980; Knauft 1985; Kelly 1993; Gérard 2017). There was always risk in moving between human and spirit worlds; those who did so could be ensnared by spirit beings, disappear into the forest and be forced to abandon their human kin. Significant cultural capital could be accrued, however, by those who successfully traversed the boundary between worlds and returned. Elsewhere in PNG, 'big men' brokered relations with people in and beyond their communities, accruing political power through mobilising the resources of others to meet aspirations of kin in marriage negotiations or intergroup conflicts (Godelier and Strathern 1991).

Colonisers, missionaries and prospectors, bringing with them knowledge of other worlds, arrived comparatively late in the land of Febi and Kubo people. Australian government patrols first arrived to document people and land in the mid-1960s, missionaries did not establish a base in the area until 1980 and, though prospecting for oil and gas in the area has been intermittent since as early as 1948, it was not until 2006 that plans for extraction took shape and multinational companies established a persistent presence in the region (Minnegal and Dwyer 2017: 53–87).

The first people brokering relations with these new worlds were themselves outsiders, sent to secure access to resources (land, labour, souls, gold, oil and gas, timber) that the state, church and corporations desired: government patrol officers, expatriate missionaries, community affairs officers sent by corporations to raise 'awareness' of development plans. While these initial incursions often were led by white men, all were accompanied by Papua New Guineans from elsewhere—policemen, evangelists, interpreters. The latter tended to be the more influential brokers in these contact situations, for they usually interacted on a more direct interpersonal level with local people. But brokers also emerged from within. Some were co-opted by the new arrivals, others by local people. And some individuals actively pursued the role of broker, attracted by the excitement and perhaps the danger entailed, as well as the possible economic and political benefits.

None of this is unique to the region where Febi and Kubo live, or even to PNG. Analyses of colonialism around the globe are replete with tales of cultural brokers and political middle-men (Shellam et al. 2016; de Jong 2018). Interest in the role of brokers faded with the end of the colonial era, as the agents and institutions of the postcolonial state were increasingly seen as key to shaping processes of social change (Lindquist 2015), and as new class dynamics and individual interests seemed to be replacing dynamics grounded in kinship and ethnicity as drivers of change (Rodman and Counts 1983; Gewertz and Errington 1999). But the neocolonialism framed by resource extraction driven by multinational corporations, by tourism with its cultural commodification, and by large-scale movements of political, social, economic and environmental refugees, together with the rise of neoliberal ideology and its reframing of the relationship between state and markets, has led to a resurgence of interest in the brokerage that these phenomena entail (Lindquist 2015; Meehan and Plonski 2017; Hönke and Müller 2018).

Whereas earlier analyses tended to presume that brokers mediated between already existing cultures, the studies emerging now see brokers as themselves active in producing, encapsulating and commodifying identities (Lindquist 2015; Minnegal and Dwyer 2017). This renewed focus moves beyond ideas of cultures as static and bounded entities, to seeing them as crystalised in and through encounter; brokers actively seek to differentiate their 'clients' from those of competitors, homogenise their client set and sell the claims of that set as distinct from those of others. Similarly, these approaches move beyond a static conceptualisation of the relationship between the 'local', 'national' or 'global', instead understanding scale as emergent and directing attention to the scale-making projects that frame encounters (Tsing 2000). This, in turn, alters the way that brokers are conceived—as mediators, who 'transform, translate, distort, and modify the meaning or the elements they are supposed to carry', rather than as an intermediary, who 'transports meaning without transformation' (Latour 2005: 39). Brokers are thus positioned at the centre of analysis, as a starting point for considering the processes that underpin production of social forms.

Again, while initial analyses focused on internal brokers as both 'exemplary' and 'exceptional' individuals (de Jong 2018), later analyses recognised that individual attributes were less important than deep-seated relations of social and economic inequality—based as much on access to education and experience as on differential control of resources—in shaping brokerage opportunities (Mosse and Lewis 2006). More recent studies, however, are concerned with the ways that brokerage may actively enhance perceptions of inequality between the fields it mediates, and construct actual inequality within those fields (James 2011; Koster and van Leynseele 2018; Bräuchler 2019: 455). While gender and generation may constrain access to brokering roles, brokerage itself may reinforce such categorical distinctions.

Recent research in Melanesian communities, too, has placed brokers centrally in analyses, with much attention to the ambiguities that frame the precarious position of these individuals and to the inequalities that they both navigate and generate through their actions (Martin 2013; Golub 2014; Schwoerer 2018). These studies, however, have tended to focus on the institutional settings within which brokers negotiate, rather than following brokers themselves through those settings.

The story we recount of one particular broker is based on conversations with him and about him with other Febi and Kubo people, over more than 20 years. We have walked with him through the bush and sat with him in houses and at feasts. While we did not follow him in person as he moved beyond his homelands to the highlands and to the capital of PNG, we have watched his appearances in newspapers and on television in his self-proclaimed role as spokesman for his people. And recently we have followed his posts on public Facebook pages, as well as those of his acolytes as some began to challenge his claim to that role.[2] Finally, information from government gazettes, public databases such as that accessed through the 'Do It Online' service of the Investment Promotion Authority (www.ipa.gov.pg), and the research tools provided by the PNGi Portal (pngiportal.org) has revealed much about connections between people and events.

The Story of a Broker

First Steps

We first met Bob Resa in October 1995. A Febi man then in his mid-30s, he was a small boy when Australian government patrols first explored the rugged landscape of his homeland in the mid and late 1960s. But the

---

2   We have been studying social change among people in this region since 1986 and, unsurprisingly, both the foci of our research and our methods have themselves changed over time. What we never expected was the ways in which international mobile phone calls and Facebook would become crucial research tools. These are sources of much of the information in this chapter.

government seldom ventured into this remote region, and made little effort to impose its influence on local people or entice them to more central locations. Bob grew up on his own land, coming to know the place and its stories. Then, in the mid-1970s, new outsiders appeared and this time Bob followed them. As a young man, he established contacts with Huli people and the Christian Brethren Church across the mountains to the northeast, and spent several years at bible school in Wewak.

In the late 1980s, Bob returned to the land of his clan. He married and, for some years, served as pastor. By 1990, he represented the community as ward councillor (*kaunsil*) in the North Koroba Rural Local-Level Government (LLG), which had offices at Koroba across the mountains to the east, in what was then Southern Highlands Province.[3] In the early 1990s, he was focal in establishing Siabi village (Figure 4.1) on the land of his fathers, in the area where exploration for petroleum had begun. He encouraged families from his own and related clans to move there, promising access to work with the exploration teams and benefits when gas was found. He dictated the layout of the village and allocated house sites. In 1995, when we visited, houses were arrayed along both sides of a straight road edged by deep ditches, paths inset with stepping-stones led to the main water-source and washing places, and some multi-storey houses had been built. All this was quite unlike our previous experiences of local house structures or hamlet designs.

The Juha area was—and still is—exceptionally isolated. There is no airstrip within three days walk from Siabi, no government services, no roads and—until very recently—almost no money. Bob gardened and hunted as others at Siabi did. But he was not the same as those others. In his earlier travels, he had learned Tok Pisin, though not English, and made valuable contacts with nationals of other language groups who were associated with missions, petroleum companies and government. In these ways, he remained connected to a world beyond Juha. As a pastor and a councillor, he drew resources and knowledge from the worlds of church

---

3   The area with which Bob is associated is, geographically, within Western Province and, politically, falls under the umbrella of Nomad LLG. However, several villages within this region are listed as being under the jurisdiction of Koroba Rural LLG. In 1990, Bob challenged patrolling census workers on the grounds that they were including villages for which he was councillor with counts for Western Province when, in fact, his people were ignored by Western Province and recognised by only Southern Highlands Province. It was more than 10 years before Bob represented his people's interests by reference to a Western Province identity.

and state into his community. And through his personal connections with men from the larger, more powerful, language groups to the north he began to venture into the world of business.

We had heard tell of Bob before we met him, in 1995, on the track between the Febi community of Siabi and the Kubo community of Suabi (Figure 4.1). A dispute had arisen and Bob was abandoning the village he had established a few years earlier. Through much of the three-day walk Bob carried his young son on his shoulders, teaching him Tok Pisin—the language, he told us, that the boy would need in the future.

For the next 12 years, Bob was based at Suabi. His wife died and he remarried, fathering three daughters. He shared ownership of a trade store, contemplated an eco-tourism venture, and in 1997 was named as director and secretary of an officially registered company.[4] He experimented with growing agarwood,[5] and accessed outside funds with which he purchased a rice mill. In the mid-2000s, he negotiated a loan to purchase a walkabout sawmill.[6] These were all attempts to benefit both the community and himself, to access opportunities that he had seen people elsewhere enjoy. His reputation as a man who got things done grew.

Becoming a Broker: Acceptance and Doubt

By the mid-2000s, plans for the PNG LNG project began to take shape. Oil Search undertook additional drilling in the mountains around Juha, operating from a base at Suabi. And through that period Bob came into his own. He was, for example, recognised as the principal landowner

---

4   PNG Investment Promotion Authority records show that the sole shareholder of that company was a Huli man who, by 2017, served as Registrar of Companies and Chairman of the Securities Commission of PNG.

5   Agarwood is a species of *Aquilaria* (also known as eagle wood, gaharu, 'gold tree' or the 'wood of the Gods') that produces a dark resinous wood in response to fungal infection. The resinous wood is used as incense and for medicinal purposes in the Middle East and Asia, and may fetch up to USD30,000 per kilogram. Local people had been selling small quantities of agarwood harvested from wild-growing trees, but Bob planted a few trees in a small plot near his house in an attempt to increase, and have greater control over, production.

6   These ventures all ultimately failed. Bob initially discussed his hopes for eco-tourism with us in 1995, but no scheme eventuated. The rice mill was not maintained, and those who had begun to grow rice abandoned their efforts when faced with a two-day walk to the nearest working mill. Oil Search had advanced the money to purchase the walk-about sawmill so that planks could be provided to floor drilling platforms, but when the company departed in 2008 the sawmill failed to attract others willing to pay the cost of hire; when it too broke down, it was locked away in a shed. In 2014, the local community health worker tried to bring the sawmill back into operation, hoping to mill local timber for a new health centre, but it had deteriorated beyond repair.

representative by Oil Search. Their community newsletter of February 2007 highlighted 'three-way cooperation' between Oil Search, the Western Province government and local landowners—the last represented by Bob. He was regarded as final arbiter on questions of ownership, made decisions with respect to employment practices at then current drilling sites, and was singled out for praise in both deflecting and making public a payment offer that was potentially corrupt (Kia and Mora 2008). In addition, Bob was recognised as an 'authority' by visiting academics who conducted social mapping and heritage studies, and as having the qualities of 'an intelligent persuasive person whose ideas and organisational abilities were highly respected'. He was seen as 'a person concerned with the welfare of the community', as someone 'whose opinion is normally accepted with silent assent in discussions' (Ernst 2008: 66). Similarly, to archaeologists who surveyed the route of a proposed pipeline along which gas from Juha would flow, Bob served as 'community liaison and translator' and, in matters concerning Febi people, was their 'major informant' (Denham et al. 2009: 4.21, 4.23).

In 2005, guided by Huli contacts, Bob registered a company under the name of Juha Development Corporation (JDC). There were 11 named directors—one from each of the then-recognised Febi clans—with Bob as chairman. This company operated as a subsidiary to the well-established Huli-based Gigira Development Corporation (GDC), providing services to petroleum companies that worked at Juha. In early 2008, in payment for those services, the Huli company directed PGK810,693.13 to the Febi company (Goldman 2009: 3–101; Minnegal and Dwyer 2017: 162). Some of that money was used to fund construction of a sawn-timber building beside the airstrip at Suabi to serve as headquarters for JDC and, perhaps, to rent to exploration companies using the strip. Some may have paid to build a two-roomed house of sawn timber and tin for a community schoolteacher.[7]

But now things started to go awry. The Gigira payment was divided among 11 people, but these were not the originally named directors. Bob received a share, but so did another man from his clan. Representatives of another five Febi clans also received shares, but six of the originally nominated Febi clans missed out entirely. And five Huli men—some connected to the parent company (GDC), others long-term advisers to Bob—received

---

7   Of money spent in the community on the two buildings, much will have been paid to Bob as hire charges for use of the sawmill.

45 per cent of the payment. By this time too, in his capacity of primary landowner representative, Bob had received more than PGK100,000 in compensation payments for environmental damage associated with land clearance at Juha drilling sites.

Now, with money in hand, Bob departed for Port Moresby, the capital city of PNG and the location of the head offices of all national government departments. Only there, he told people, would he have access to politicians, bureaucrats and officers of petroleum companies. Only through direct contact with these powerful 'others' could he ensure access for his Febi compatriots to financial benefits from the PNG LNG project—to business development grants, infrastructure grants and, eventually, royalties and equity. In Port Moresby, he would be able to monitor bureaucratic and legal processes, and intervene if needed so that they 'would eventually receive what was rightly theirs' (Minnegal and Dwyer 2017: 152).

Initially, at least, Bob's efforts to differentiate and promote the interests of his people appeared to be succeeding. He was acknowledged by Department of Petroleum and Energy (DPE) as a major representative of landowners associated with the Juha gas field and, in this capacity, invited to participate in a May 2009 meeting that was convened to negotiate an overarching agreement between the state, provincial and local governments and landowners concerning the future distribution of benefits through the entire PNG LNG project area (GPNG 2009; Filer 2019: 32–35). In fact, Bob chose not to go and was able to stop most other nominated Juha representatives from attending. At this time, he was promoting a view that Juha should be operated as a Western Province, stand-alone venture, physically and economically disassociated from the rest of the PNG LNG project. His campaign attracted media attention, and Bob began to develop a wider profile as spokesman for his people. There was no chance, however, that separate development would be viable; the Juha gas field made only a minor contribution to the whole. By November 2009, Bob had put these plans aside and was a major contributor at a forum, specifically concerned with distribution of future benefits from Juha's association with the PNG LNG project, that was convened, initially, at Suabi. Huli participants, seeking to establish their own status as beneficiary landowners, complained that the venue was too muddy, facilities and food were inadequate, women came to the meetings and there was risk of sorcery with Bob always 'looking at' them. In response to their intense lobbying the forum was relocated to Moro, a long-established base for Oil Search, near Lake Kutubu in Southern Highlands Province (Minnegal and Dwyer 2017: 83, n. 18). Initially, Bob's representations on behalf of his people prevailed. The draft agreement presented at Moro stated that beneficiary landowners for the Juha area would comprise 12 Febi clans together with other Western and Hela Province clans 'as invited by the Febi clans'.[8] The Febi clans were to receive 90 per cent of Juha-derived royalties. In the end, however, the meeting was judged to have failed; in March 2010 the Minister for Petroleum and Energy signed an interim determination that reduced the proposed share to Febi clans from 90 to 50 per cent of the total; and 10 years later, despite two attempts to vet landowners, no final determination of beneficiary landowners has been gazetted.

In the course of these negotiations, the idea of 'Febi' as a collective identity, on behalf of whom one could speak and enter into a formal agreement, was becoming more concrete. Much of this shift may be attributed to Bob's personality and actions. In the years that followed, he encouraged Febi people to channel all applications for business and infrastructure development grants through him. Many did so and, though some were disgruntled by his failure to share in expected ways, most still felt he was a good man, working on their behalf. They certainly considered him better placed than anyone else to mediate relationships with government and petroleum companies. In acknowledgement of this assigned status, Bob began to speak of himself as 'Chief' or, indeed, 'Paramount Chief' of the Febi people. In doing so, he both reified Febi as a distinct collectivity bound by common interests and set himself apart from, and above, other Febi in a manner that was entirely alien to the contingent inequalities that had, heretofore, prevailed among people of the Strickland-Bosavi region (cf. Kelly 1993).

Throughout this period, Bob's relationships with Huli men were ambiguous. Some of those men were persistent in promoting their own claims to ownership with respect to Juha. Others were not only Bob's intellectual advisers, but also became his sponsors in establishing necessary contacts, his monetary benefactors in providing long-term loans and, as is standard Huli practice, his affinal kin in encouraging him to abandon the mother of his daughters and marry a Huli woman. Guided by these men, Bob pursued funds in the name of Febi people. Details are fragmentary,

---

8   Juha Petroleum Retention Licence 2 Licence Benefits Sharing Agreement 2009, copy held by authors.

but, at the least, he accessed more than 2 million kina and was supported for two short visits to Australia. In Port Moresby, he bought a house, acquired office space, and supported several young male acolytes who could read and write on his behalf (Minnegal and Dwyer 2017: 152–53); the latter were mainly Febi from clans other than his own whom he sponsored to study in the city. He hired legal and financial advisers. And he ensured that his daughters continued their high school education.

None of the money Bob now secured, however, was directed to developments within the territories of Febi or Kubo people. People living there began to wonder what had happened to him. In 2011, a message addressed to Bob was scrawled on the wall of the now-disused JDC building: it asked where he was and asserted 'someone wants to marry your daughter'. Bob's relationships to the Febi people he purportedly represented were beginning to dissolve.

In December 2012, Bob briefly returned to Suabi. He was guest of honour at a major community feast, a feast that was planned by senior residents with the deliberate aim of attracting people who had moved elsewhere and appeared to have 'forgotten' their origins. Bob accepted the 'invitation', but insisted that the feast be an occasion for cultural revival. He came by chartered plane, accompanied by his acolytes. From a podium built so he could oversee the cultural performances he had demanded—mock raids, costumed dancing, hilarious performances by men disguised as ogres—Bob addressed his people. 'He spoke of all he had done, and was doing, for the community, hinted at planned autonomy for the Juha area', and talked of the desirability of electing a 'Juha president' (Minnegal and Dwyer 2017: 153).

Bob's audience was not satisfied. They challenged him, drawing attention to the fact that they had seen no benefit from the money he had accessed. That money merely fed his life in town, generating inequities in access to resources and in quality of life. He had abandoned his place and his people, they declared. After Bob returned to Port Moresby, people asserted that if he came again he would have to rent accommodation or sleep outside under coconut palms that he himself had planted years earlier. Later, in 2014, when we censused the village, even his closest kin in the village insisted that he should not be listed as an 'absentee resident', he was no longer to be accorded status of any sort as a 'resident'.

To Bob, however, these accusations were unwarranted. Local people did not understand the importance of what he was doing in Moresby, he told us; they did not appreciate his persistent efforts on their behalf.

The Fall

In May 2014, the PNG LNG project shipped its first load of gas out of the country. There have now been more than 500 shipments to Japan, China and Korea. None of the gas came from Juha, but, under the 2009 umbrella benefit-sharing agreement, Juha landowners were entitled to receive 2.02 per cent of royalties on those sales.[9] By late 2019, however, no royalties had been paid to any gas field landowners, even to owners of the fields from which gas has been taken.[10]

At all petroleum licence areas there are huge problems entailed in identifying legitimate landowners (Koim and Howes 2016; Filer 2019). The task of assessing claims falls to the DPE and/or the judiciary. The final decision, however, is that of the Minister for Petroleum and Energy (now the Minister for Petroleum), who is not obliged to accede to the advice of either the legally required social mapping reports or officers of his own department. As a consequence, the potential for lobbying and 'clientelism' is great. And thus the importance of well-placed brokers is also great.

Of recent years, this is where Bob has devoted his efforts. But, so often, those efforts have been thwarted. An interim ministerial determination of Juha beneficiaries in 2010 was put aside. A process of clan vetting in November 2013 was challenged and judged to be inadequate. The process was scheduled to recommence late in 2017. Bob and others assembled at the proposed venue but promised funds failed to materialise, DPE staff did not arrive, and nothing was accomplished. The process almost got off the ground in mid-February 2018 but, again, at the last minute those plans were put on hold; a massive earthquake intruded, the epicentre

---

9   According to the benefit-sharing agreement negotiated at the start of the PNG LNG project, as soon as gas is sold from any one of the participating gas fields all licence areas are to receive a share of payments proportional to their anticipated contribution of gas to the project as a whole. Juha landowners are thus eligible to receive 2.02 per cent of royalties even though production has not yet commenced there and is unlikely to do so in the next decade.

10   Owners of the land on which the LNG processing plant at Caution Bay, near Port Moresby, was built have received royalty payments.

close to the Hides gas conditioning plant and the impact reverberating throughout the land of Febi and Kubo people (Dwyer and Minnegal 2018; Main 2018; Zahirovic et al. 2018).

In the course of these promises and frustrations, Bob continued to seek and report connections with people he judged to be in a position to help. On his behalf, and often in his name, others promoted his cause through social media, particularly posts to a widely followed Western Province forum. In one 2017 post, where he was named as 'Chief of Chiefs ... of JUHA PDL9' he is 'joint hands together' with 'HON. Dr. Fabian Pok, Minister for DPE' as they planned a 'way forward in clan vetting ... for Juha PDL9 projects area'. In another post, from January 2018, he is named as 'paramount chief of Juha PDL 09'. The post shows him accompanied by a Huli man who, six years earlier, had acted as consultant in promoting the aims of one of Bob's companies and, post-earthquake, was appointed deputy provincial administrator for Hela Province; together, they are about to present a proposal to the government clan-vetting team led by DPE Vice Minister Manessah Makiba. And, in a May 2018 TVWAN video clip, Bob appears with purported landowners of the Angore petroleum licence area, one month before they burned ExxonMobil equipment at the Angore wellheads. The man speaking for Bob on this occasion is brother to the Huli man who sponsored Bob's 2016 trip to Australia.

None of these efforts came to fruition. Bob aged. He became despondent. A Facebook post from February 2018—we presume it was 'written on his behalf'—stated:

> I am in Bomana jail. I am the Chief of Juha PDL9. I do not know when I will get out of jail and go to my place. I would like the Governors of Hela and Western Provinces to release me so that I can go to my place now. [Our translation from the original Tok Pisin.]

The allusion to jail was metaphoric. Bob was tired of waiting, exhausted by his failed efforts, unsure why the process was, once again, stalled. He was tempted to nostalgic recall of the place he once knew as home. But who, or what, was keeping him 'in jail'?

Competition and Marginalisation

In late June 2018, quite suddenly, it seemed that there was a turn for the better. DPE officers arrived at Kiunga, a river town 100 kilometres west of Suabi. The Port Moresby–based Juha claimants—including Bob—flew in. DPE rustled up others who worked in Kiunga, lived in Kiunga or happened to be visiting Kiunga. Huli claimants from Koroba had access to funds and were well represented. But most Juha clans were unrepresented; no funds were available to bring rural people to Kiunga, and the only means of travel from Suabi were by air or a four- to five-day walk.[11]

There were multiple meetings: meetings to name the beneficiary clans and agree to their respective shares; meetings to elect representatives to an umbrella association and an umbrella company; a meeting to schedule another meeting at which Incorporated Land Groups (ILGs)[12] would be established to enable receipt of payments; and, in the absence of DPE officers, 'unity' meetings to devise strategies for future negotiations.

Despite all this talk, there was no final resolution. The division of royalties to Huli claimants was left to future negotiations in Port Moresby; these people had exceeded the tolerance of DPE officers in listing the names of several hundred clans—most freshly created—as deserving beneficiaries. Nothing was done with regard to setting up ILGs. That remained the state of play more than three years later with a final ministerial determination of Juha beneficiaries still pending.

But there were other reasons for dissatisfaction. As Wisu Miago—another, younger, Febi man at these meetings who had been one of Bob's early 'acolytes'—commented in a public Facebook post, the process started late and ran over time. The DPE officers, he asserted, lived in luxury but extended no courtesies or allowances to Juha landowners who, as a result, were stranded in Kiunga, living on credit, waiting for the Fly River Provincial Government to fund return flights to Port Moresby. Other posts reinforced and elaborated these observations.

---

11   Unable to attend the negotiations themselves, people at Suabi queued on the hill behind the village each evening to hear news of progress via mobile phone. We too, back in Australia, waited avidly for news via phone from our contacts in Suabi. Few were entirely happy with what was reported, but told us that it was time for dispute to end. They hoped that with agreement reached over identification of landowners, even if that was not accurate, at least some money would reach the community.
12   For a discussion of Incorporated Land Groups, the processes entailed in setting these up, and ambiguities about their status as the legal entities to which benefits from LNG projects are to be paid, see Minnegal et al. (2015).

Wisu had other things in mind, however. He had been elected chair of the proposed new Umbrella Company, while Bob received a less significant position as chair of the Umbrella Association. Wisu, it seems, was quietly out-manoeuvring Bob, his one-time mentor.

In August 2018, still stranded in Kiunga, Wisu 'shared a memory' on Facebook. A year earlier, he had returned to Suabi, joining with kin to celebrate the opening of a new classroom for the Juha Elementary School. His 'leadership' was noted by Huli Facebook friends: 'you can lead Juha landowners' they declared, remarking that he attended to the 'grassroots'. In acknowledging the compliments, Wisu responded that there were some people who lived a 'luxurious life' in Port Moresby and forgot 'priorities'. He did not name anyone—that is not Febi or Kubo practice—but his allusion was unambiguous to people in the know.

A few days later, Wisu posted a photograph of a bowl of breadfruit nuts. He wrote of the hardship and sacrifices entailed in leadership and commented that feeding on the breadfruit would equip him for taking up the roles and responsibilities required of a leader. Stranded at Kiunga, living on credit, Wisu and others were reduced to harvesting breadfruit for food. Breadfruit, however, is seen as bush food; it grows like a weed along the edges of roads and needs no human labour to produce. For Kubo and Febi, to offer such food to guests would be shameful, indicating you were not able, or willing, to invest in a relationship with them (Minnegal and Dwyer 2001: 282). Wisu turned this 'shame' to good effect, however. His Facebook post revealed that he would accept the 'hardship' entailed in making 'sacrifices' on behalf of others. He had shown himself willing to assume the responsibilities, and associated pains, of a true leader.

Through late 2018 and 2019, Wisu's prominence in Juha-related negotiations has continued to grow, as Bob's visibility has declined. Like Bob, Wisu is based in Port Moresby but, unlike Bob, he has no history of demonstrated 'good works' in his home community.[13] Further, his audience is reached via social media that few Febi or Kubo people can routinely access and his support comes from Huli friends and sponsors whose motives are regarded as doubtful by most Febi and Kubo people. Bob may have overreached with his claim to be 'chief of the Febi', but

13   The new elementary school classroom, for the opening of which he had returned to Suabi, was built of bush materials by local people while Wisu was in Port Moresby. He had invested neither money or labour in its construction.

his attempts to broker relationships with government and company were undoubtedly endorsed—at least initially—by those people and were recognised by the academic and administrative authorities with whom he interacted. Further, Bob was of Wuo clan, unambiguously associated with the Juha wellheads, whereas Wisu's natal affiliations are with another clan whose land does not encompass any of the four wellheads. Wisu is courting support and authority from among people who live beyond, or have ventured beyond, Febi land, and not directly from people who remain in place. Within that wider constituency, the demands for, and the reach of, mediation have grown. Through early 2019, Wisu was acting as spokesman not just for Febi people but for their Siali neighbours, as the latter sought recognition as beneficiary landowners of PDL 7 and the Juha pipeline route. The stage is shifting. And the networks from which power is perceived to derive, the sources of authority, too, are shifting. Culturally specific measures of worth and capacity remain relevant, but those who judge these are now more likely to be from elsewhere.

Discussion

By the mid-2000s Bob Resa had assumed, and was accorded, a leadership role among Febi people. He was experienced, knowledgeable, persuasive. He was concerned with the welfare of the community. Among people of the Strickland-Bosavi area these were qualities that gave men authority. But there was much more to it than that. Bob, it seemed, was able to venture into worlds beyond the horizons of everyday life for most Febi and Kubo, and establish relationships with those he met there. In this, he resembled the *save* men—men of knowledge—who had in the past mediated relations with the spirit world. But negotiating with spirits was not a task to be taken on lightly. It was a task for men with the strength to resist the lure of life with their spirit affines, able to control their own desires and behaviours, and aware of the risks entailed (Gérard 2017). Few who ventured into that other world managed to return, with or without the resources they had sought for their kin. Those who were not vigilant—who, for example, ate the food of the spirits by mistake—would become spirits themselves, lost to their human kin.

Gérard (2017) has argued that among Febi, since colonisation and, particularly, exposure to Christian teachings and the arrival of men seeking petroleum resources, there has been a shift in emphasis with respect to desired resources; it is money now, rather than game, that people seek. As before, however, some men are more able than others to move between worlds, to build relationships with the outsiders, the inhabitants of those other worlds, who are the ultimate arbiters of access to money and well-being: Huli, government, God and petroleum companies. Bob Resa was such a man; a post-contact analogue of a 'spirit medium'.

For some years now, however, Bob has shown signs of being captured by the 'other': he has a wife from outside, lives apart from his own people in the domain of the other, has failed to bring back the resources he accessed there and, as evidenced by becoming fat, is 'greedy' for city life. Some people would like to reject him. Others are uncertain who could replace him. Many remain anxious that he may be spiritually powerful and dangerous. They fear him, even as they envy him.

Bob Resa is not alone in moving to Moresby. Others, like Wisu, have followed, in part seeking to emulate Bob but also seeking to succeed, where Bob has failed, in bringing wealth from outside back to the community. So far, however, despite Wisu's efforts to imply otherwise, it seems no one has managed to cross to the 'other side' and then return.[14] A new way of being Febi is emerging. And it may soon prove to be the dominant mode of being Febi. At present, as it comes into being, it generates inequality. And, in different ways, both for those at home and those elsewhere, it generates senses of loss; for those at home, loss of what might have been if the spirit of egalitarianism had prevailed, and for those who departed, loss of what once was but, increasingly, is recalled only as stories, not as lived experience.

For people like Bob, who has left, and can probably never return until he dies, we must appreciate that, despite the outcomes of his behaviour, he himself continues to imagine that his actions are honourable, that to achieve desired ends for the community he serves he must remain in the metaphorical jail that is Port Moresby.

---

14 Like local people, however, we retain hope that a 'save man' may one day return with the keys to well-being in the community. The young man elected as councillor for Suabi ward in 2019 had returned to the village at the end of 2014 after post-secondary studies in Port Moresby. He is the eldest son of a highly respected pastor at Suabi who died in 2010. He has chosen to remain in the village, where he advises people to stop waiting for royalties that may never be paid, and encourages them to establish small agricultural businesses to supply local demand (fish, chickens, rice) and, perhaps, export (cocoa, vanilla, agarwood). In 2019 a new school building of the sort people had long desired—steel-framed, with tin roof and glass windows—was constructed at Suabi, funded by the provincial government through the PNG LNG Development Levy.

Conclusion

With specific attention to South Africa, James (2011: 318) has argued that brokers do not 'merely negotiate' between the 'fixed positionalities' of, on the one hand, 'people' and, on the other, the 'state' or the 'market'. Rather, she wrote:

> they embody and bring into being socio-economic positions and identities. They blend together the egalitarianism and rights-based character of post-liberation society with the hierarchy of re-emerging traditional authority.

Her general point seems appropriate to the present case study. Bob Resa, as broker, is deeply implicated in what we see as epistemological and ontological shifts that are transforming Febi and Kubo into 'new kinds of people' (Minnegal and Dwyer 2017). To paraphrase Gewertz and Errington (2016: 350), in years to come 'the Febi will still be there and they will still be Febi, yet they will definitely and fundamentally not be their (grand)father's Febi'.

But James's processual point about brokers seems not to apply to the Febi case. What we see happening is almost the reverse: a blending, almost a submerging, of an egalitarianism implicit in traditional Febi society with expressions of hierarchy—of status, factionalism and possessive individualism—perceived to be operative in modern, neoliberal society.

Acknowledgements

We thank Febi and Kubo people for their hospitality and teaching since 1986, and Anaïs Gérard for sharing information and insights about Febi people. An earlier version of this paper was presented at the session 'Resource extraction and dealing with inequality in the Pacific', European Society for Oceanists 2018 Conference, Cambridge, UK. Our recent research in PNG has been supported by an Australian Research Council Discovery Grant (DP120102162), and the insights about brokers shared in this paper form the basis for another ARC-funded project (DP220101633).

References

Bainton, N.A. and G. Banks, 2018. 'Land and Access: A Framework for Analysing Mining, Migration and Development in Melanesia.' *Sustainable Development* 26(5): 450–460. doi.org/10.1002/sd.1890

Bräuchler, B., 2019. 'Brokerage, Creativity and Space: Protest Culture in Indonesia.' *Journal of Intercultural Studies* 40(4): 451–468. doi.org/10.1080/07256868. 2019.1628721

Cox, J., 2018. *Fast Money Schemes: Hope and Deception in Papua New Guinea.* Bloomington: Indiana University Press. doi.org/10.2307/j.ctv6mtfjm

de Jong, S., 2018. 'Brokerage and Transnationalism: Present and Past Intermediaries, Social Mobility, and Mixed Loyalties.' *Identities* 25(5): 610–628. doi.org/10.10 80/1070289X.2018.1515778

Denham, T., A. Bedingfield and U. Gilad, 2009. 'Juha to Hides.' In L. Goldman (ed.), *Papua New Guinea Liquefied Natural Gas Project: Social Impact Assessment 2008.* Report to ExxonMobil Corporation.

Dwyer, P.D. and M. Minnegal, 1992. 'Ecology and Community Dynamics of Kubo People in the Tropical Lowlands of Papua New Guinea.' *Human Ecology* 20(1): 21–55. doi.org/10.1007/BF00889695

———, 2018. 'Refugees on Their Own Land: Edolo People, Land and Earthquakes.' *EnviroSociety* blog, 9 June. Viewed 18 May 2020 at: www.enviro society.org/2018/06/refugees-on-their-own-land-edolo-people-land-and-earthquakes

Ernst, T. 2008. 'Full-Scale Social Mapping and Landowner Identification Study of PRL02.' Unpublished Report to ExxonMobil Corporation (Copy held by authors).

Filer, C., 2019. 'Methods in the Madness: The "Landowner Problem" in the PNG LNG Project.' Development Policy Centre Discussion Paper 76. doi.org/ 10.2139/ssrn.3332826

Foster, R.J. and H.A. Horst (eds), 2018. *The Moral Economy of Mobile Phones: Pacific Islands Perspectives.* Canberra: ANU Press. doi.org/10.22459/MEMP. 05.2018

Gérard, A., 2017. Procuring Game, Procuring Money: Dilemmas of Relationality with Outsiders among Febi People, Western Province, Papua New Guinea. Melbourne: University of Melbourne (Unpublished PhD thesis).

Gewertz, D.B. and F.K. Errington, 1999. *Emerging Class in Papua New Guinea: The Telling of Difference*. Cambridge: Cambridge University Press. doi.org/10.1017/CBO9780511606120

———, 2016. 'Retelling Chambri Lives: Ontological Bricolage.' *The Contemporary Pacific* 28: 347–381. doi.org/10.1353/cp.2016.0031

Godelier, M. and M. Strathern (eds), 1991. *Big Men and Great Men: Personifications of Power in Melanesia*. Cambridge: Cambridge University Press.

Goldman, L. (ed.), 2009. *Papua New Guinea Liquefied Natural Gas Project: Social Impact Assessment 2008*. Report to ExxonMobil Corporation.

Golub, A., 2014. *Leviathans at the Gold Mine: Creating Indigenous and Corporate Actors in Papua New Guinea*. Durham: Duke University Press. doi.org/10.1515/9780822377399

Goodhand, J., B. Klem and O. Walton, 2016. 'Mediating the Margins: The Role of Brokers and the Eastern Provincial Council in Sri Lanka's Post-War Transition.' *Third World Thematics: A TWQ Journal* 1(6): 817–836. doi.org/10.1080/2380 2014.2016.1302816

GPNG (Government of Papua New Guinea), 2009. 'PNG LNG Project: Umbrella Benefits Sharing Agreement.' Viewed 18 November 2015 at: actnowpng.org/sites/default/files/UBSA.pdf

Hönke, J. and M.-M. Müller, 2018. 'Brokerage, Intermediation, Translation.' In T.A. Börzel, T. Risse and A. Draude (eds), *The Oxford Handbook of Governance and Limited Statehood*. Oxford: Oxford University Press. doi.org/10.1093/oxfordhb/9780198797203.013.16

James, D., 2011. 'The Return of the Broker: Consensus, Hierarchy, and Choice in South African Land Reform.' *Journal of the Royal Anthropological Institute* 17(2): 318–338. doi.org/10.1111/j.1467-9655.2011.01682.x

Kelly, R.C., 1993. *Constructing Inequality: The Fabrication of a Hierarchy of Virtue among the Etoro*. Ann Arbor: University of Michigan Press.

Kia, J, and S. Mora, 2008. 'Juha 4 & 5 CA Close-Out Report – PRL02. Report to Oil Search Ltd, 13 November 2008' (Copy held by authors).

Knauft, B.M., 1985. *Good Company and Violence: Sorcery and Social Action in a Lowland New Guinea Society*. Berkeley: University of California Press.

————, 1998. 'How the World Turns Upside Down: Changing Geographies of Power and Spiritual Influence among the Gebusi.' In L.R. Goldman and C. Ballard (eds), *Fluid Ontologies: Myth, Ritual and Philosophy in the Highlands of Papua New Guinea.* London: Bergin and Garvey.

Koim, S. and S. Howes. 2016. 'PNG LNG Landowner Royalties – Why so Long?' *DevPolicy* blog, 16 December. Viewed 18 May 2020 at: www.devpolicy.org/png-lng-landowner-royalties-long-20161216/.

Koster, M. and Y. van Leynseele, 2018. 'Brokers as Assemblers: Studying Development through the Lens of Brokerage.' *Ethnos* 83(5): 803–813. doi.org/10.1080/00141844.2017.1362451

Latour, B., 2005. *Reassembling the Social: An Introduction to Actor-Network-Theory.* Oxford: Oxford University Press.

Lindquist, J., 2015. 'Brokers and Brokerage, Anthropology of.' In J.D. Wright (ed.), *International Encyclopedia of Social and Behavioral Science* (2nd edition). Amsterdam: Elsevier. doi.org/10.1016/B978-0-08-097086-8.12178-6

Main, M., 2018. 'How PNG LNG is Shaking Up the Earthquake.' *EnviroSociety* blog, 28 March. Viewed 18 May 2020 at: www.envirosociety.org/2018/03/michael-main-how-png-lng-is-shaking-up-the-earthquake.

Martin, K., 2013. *The Death of the Big Men and the Rise of the Big Shots: Custom and Conflict in East New Britain.* New York: Berghahn Books.

Meehan, P. and S. Plonski, 2017. *Brokering the Margins: A Review of Concepts and Methods* (Borderlands, Brokers and Peacebuilding Project, Working Paper No. 1). London: SOAS Eprints.

Minnegal, M. and P.D. Dwyer, 2001. 'Intensification, Complexity and Evolution: Insights from the Strickland-Bosavi Region.' *Asia Pacific Viewpoint* 42(2/3): 269–285. doi.org/10.1111/1467-8373.00149

————, 2017. *Navigating the Future: An Ethnography of Change in Papua New Guinea.* Canberra: ANU Press (Asia-Pacific Environment Monographs). doi.org/10.22459/NTF.06.2017

Minnegal, M., S. Lefort and P.D. Dwyer, 2015. 'Reshaping the Social: A Comparison of Fasu and Kubo-Febi Approaches to Incorporating Land Groups.' *The Asia Pacific Journal of Anthropology* 16(5): 496–513. doi.org/10.1080/14442213.2015.1085078

Mosse, D. and D. Lewis, 2006. 'Theoretical Approaches to Brokerage and Translation in Development.' In D. Lewis and D. Mosse (eds), *Development Brokers and Translators.* Kumarian, Bloomfield: Kumarian Press.

Robbins, J., 2013. 'Beyond the Suffering Subject: Toward an Anthropology of the Good.' *Journal of the Royal Anthropological Institute* 19: 447–462. doi.org/10.1111/1467-9655.12044

Rodman, W.L. and D.A. Counts (eds), 1983. *Middlemen and Brokers in Oceania.* Washington DC: University Press of America (ASAO Monograph No. 9).

Schieffelin, E.L., 1976. *The Sorrow of the Lonely and the Burning of the Dancers.* New York: St Martin's Press.

———, 1977. 'The Unseen Influence: Tranced Mediums as Historical Innovators.' *Journal de la Société des Océanistes* 33: 169–178. doi.org/10.3406/jso.1977.2954

Schwoerer, T., 2018. '*Mipela Makim Gavman*: Unofficial Village Courts and Local Perceptions of Order in the Eastern Highlands of Papua New Guinea.' *Anthropological Forum* 28(4): 342–358. doi.org/10.1080/00664677.2018.1541786

Severs, E. and S. de Jong, 2018. 'Preferable Minority Representatives: Brokerage and Betrayal.' *PS: Political Science & Politics* 51(2): 345–350. doi.org/10.1017/S1049096517002499

Shellam, T., M. Nugent, S. Konishi and A. Cadzow (eds), 2016. *Brokers and Boundaries: Colonial Exploration in Indigenous Territory.* Canberra: ANU Press. doi.org/10.22459/BB.04.2016

Sørum, A., 1980. 'In Search of the Lost Soul: Bedamini Spirit Seances and Curing Rites.' *Oceania* 50: 273–296. doi.org/10.1002/j.1834-4461.1980.tb01410.x

Tsing, A., 2000. 'The Global Situation.' *Cultural Anthropology* 15(3): 327–360. doi.org/10.1525/can.2000.15.3.327

Weiner, J.F. (ed.), 1988. *Mountain Papuans: Historical and Comparative Perspectives from New Guinea Fringe Highlands Societies.* Ann Arbor: University of Michigan Press. doi.org/10.3998/mpub.9552

Zahirovic, S., G. Brocard, J. Connell and R. Beucher, 2018. 'Aftershocks Hit Papua New Guinea as it Recovers from a Remote Major Earthquake.' *The Conversation*, 9 April. Viewed 18 May 2020 at: www.theconversation.com/aftershocks-hit-papua-new-guinea-as-it-recovers-from-a-remote-major-earthquake-94176

5

'Em i Stap Bilong En Yet': Not-Sharing, Social Inequalities and Changing Ethical Life Among Wampar

Bettina Beer

Introduction

This chapter will consider changes in the ethical dimensions of collective life among Wampar, Markham Valley, Papua New Guinea (PNG), based on discussions of 'thick ethical concepts' such as 'stinginess', 'shame' or 'gossip' recorded in the 1970s, as well as ethnographic vignettes from fieldwork in 2013 and 2017/18.[1] I focus on different economic activities: on transfers of food, money and consumer goods including sharing, distribution and perceived violations of reciprocity—themes that have been discussed in the literature on Melanesia under the keyword of 'moral equivalence' (Burridge 1969; Barker 2007). More recently, ethics has become a 'hot topic' in anthropology (cf. Mattingly and Throop 2018), with diverse theoretical approaches (Mahmood 2012; Laidlaw 2014; Mattingly 2014; Keane 2016) and empirically based contributions, out of which only a few focus on morality in Melanesia (e.g. Read 1955;

1   For feedback and discussion on the topic of this paper, I thank Don Gardner, Doris Bacalzo, Willem Church and Tobias Schwoerer; participants of colloquia at the Departments of Ethnology in Hamburg, and Lucerne; and of the workshops on social inequalities at the 2017 and 2019 conferences of the Association for Anthropology in Oceania.

Barker 2007; Robbins 2013; Busse and Sharp 2019). In this chapter, I will try to link the ethnography of current discourses on ethics and social inequality among Wampar-speakers in the Markham Valley to recent anthropological debates.

In the context of anticipated large-scale resource extraction, expectations of upward mobility and participation in the global economy increase drastically in the local population. Wampar experience changes resulting from immigration, urban growth and other effects on their social and physical environment. During the prospecting phase of mining, subcontractors and companies selling machinery, fences, catering and logistics to mining companies open opportunities for some local families and not for others. The middle-class lifestyle known from adverts, movies and visits to town seems to become available to Wampar villagers. The prospecting phase often takes decades and can produce greater changes to local social relations and, perhaps, to the environment, than actual resource extraction. The effects of 'pre-mining', as well as mining itself, connect issues described in this chapter to other chapters in this volume describing situations closer to mining (Church, Chapter 3) or even those remote from it (Knauft, Chapter 6).

Growing economic inequality, increased consumption of industrial products, and the conspicuous presence of global capital is hard to miss in the coastal area around Lae, PNG's second city, which has been exposed to colonial and missionary presence for a century (Fischer 1992; Beer and Church 2019). Early ethnography (Fischer 1975) indicates that social relations among Wampar at that time could be characterised as emphasising equivalence in a Burridgean sense (Burridge 1969; Knauft 2007). This involved favouring sister exchange in marriage, or delayed exchange in bridewealth (often after the birth of one or more children), and a stress on sharing. Negotiations of such relations in the growing presence of global capitalism have become a topic of discussion among Wampar, as among many (even most) other ethnic groups in PNG. Today, sister exchange has been given up (Beer 2015), and bridewealth negotiations have become more flexible and pragmatic, at least partly due to the interethnic situation of many married couples. However, underlying ethical assumptions about sharing as formulated in early Wampar ethnography are still evident in the context of growing social inequalities. Although now embedded in Christian morality, non-sharing remains negatively evaluated and subject to social sanctions such as gossip and sorcery accusations.

In the literature on PNG, the desire for 'development' and 'modernity' has been described and discussed in detail. Among Wampar over the last 20 years, positions vis-à-vis large-scale capital-intensive projects have been diverse and heterogeneous: some criticise the loss of commitment to close-knit kin ties and the community, as well as to land and gardening as a subsistence economy. In this context 'greed', 'stinginess' and the avoidance of sharing are discussed. These discourses on the state of social relations are not new, as I will show, but they have become more frequent and urgent as Wampar, positioned differently in emerging social hierarchies, reflect on social relations and their positions within these hierarchies.

Stinginess, Greed and Gossip

Sensitivity to transfers of valued items between people is a human universal (cf. Antweiler 2016: 110ff; Keane 2016: 15), and interpersonal as well as intergroup transfers are always likely to reflect existing relations and to affect their ongoing trajectory as recipients and givers negotiate the significance of the transfer and its social context.

Transfers, then, cannot be of one piece: in addition to variations in the quantitative dimensions of the transfer as a dimension of their significance (relative to a socio-economic context, social norms), the motivations of the transferor and the reactions of the transferee, as well as the wider social ramifications of these, make for differences between transfers. Sharing, for example, does not usually involve a concern for reciprocity, while reciprocal relations need entail no sharing, beyond a mutual concern to service and continue the relationships involved.

Recently, scholars (Woodburn 1998; Widlok 2013; Schnegg 2015) have expressed some concern with anthropology's tendency to focus on reciprocal relations and the values associated with them as the most significant social form. Here I focus on Wampar responses to transfers, as these relate to who shares what with whom, with a view to showing how the increased circulation of money and commodities has affected evaluations of folks relative to an understanding of Wampar traditions in the context of increasing social inequalities. With more money circulating and the significance it has for life chances, problems of free-riding have become more apparent and violations of values are frequently discussed.

Early Wampar texts portray a person who does not share with others, and is thus categorised as greedy and/or stingy, as becoming the subject of gossip that, rightly, occasions feelings of shame. Wampar generally try to avoid a reputation for stinginess through generosity, but often suspect or accuse others of this failing. In this section, I will discuss in detail talk involving ethical concepts such as stinginess and greed, and the gossip it promotes, as couched in the Wampar language, since some do not have an equivalent in Tok Pisin (TP, the most widespread lingua franca in PNG). If there are such equivalents, for example on 'gossip' (*tok baksait* in TP), I will discuss them too because they are essential to emerging national discourses. The terms, and discussions of them, that I present were given to Hans Fischer by his interlocutors in the 1970s. The terms are still in use and I have heard them in conversations, conflicts and diverse everyday social interactions during fieldwork from the late 1990s to 2018. I present my own material in the present tense, despite the span over which it was collected, but do so only for the sake of the reader—it does not mean my interlocutors in Gabsongkeg village are people 'frozen in time', as will become clear in the following section on global entanglements and their consequences.

The Wampar make use of the sense of smell to talk about stinginess. *Muteran* means 'to smell' or 'to stink', as well as 'to be stingy with something': *Mpi ongan emar, imut*, 'If a pig dies, it stinks', or *Ngaeng o kai gea imut inin en gaen*, 'That man is stingy with banana'.[2] Stinginess encompasses a wide range of behaviours, including the demand that somebody asks before taking food. The use of *muteran* connotes that stinginess is like a strong, offensive stench as described in the following texts, which Hans Fischer recorded in the 1970s.[3]

> Stingy [*muteran*] man and stingy woman do not let others take their things, like their banana, coconuts, betel nuts, betel pepper, their possessions, their sugar cane, or their garden products. If somebody takes something from them, they grumble: 'Who told you that you could take my betel nuts, pick my betel pepper or break my sugar cane? You have to ask me first, and then you can

2   *Imut* is specifically used for decomposing human and animal corpse but not for general unpleasant body odours of living organisms, which are labelled *renen ferentseng* or *ufin* (*funufineran*, emanating a bad smell, *rain ufin*, fart).
3   Unless otherwise indicated, the following quotes are from unpublished transcripts of texts recorded by Hans Fischer in 1976. His interlocutors were Gufose–Imanuel (1935–?) and Kupik–Emonteng (1915–1990).

128

take it.' If a woman has good manners, and things are taken from her, she'll not say anything at all. If she hears about someone taking something, she'll just say: 'It's just one thing. It's everyone's thing. It's not something for only sisters or brothers; it's for everyone.'

The stingy person, they imply, acquires a reputation that follows them around like a bad smell. The use of *imut* also alludes to the fact that having something and not sharing it is in many cases difficult to hide. The vocabulary for odours of highly desirable foods that should be shared is more sophisticated than in English or German. The delicious smell of roasted meat (or other roasted food) *ntsedz* is, for example, a specific smell term. It describes smells that can be perceived from a great distance, revealing the preparation of highly appreciated food. Therefore, one either must share with household members and kin, or roast meat or fish in the garden, or at night:

> There is a woman who is a very stingy woman. She does not even give food or anything else to her parents-in-law. When she goes fishing and takes the catch home, she does not unpack the fish but keeps them (in her bamboo tube). It's only when her in-laws are asleep that she takes everything out. She unpacks the fish and cooks it while everyone is asleep so that her parents-in-law do not see anything. She also only eats by herself. She puts the leftovers away, and on the next day, when her parents-in-law have gone away, she brings out the food, cooks it for her children and eats with them or with her husband. When she prepares food, she does not give anything to her parents-in-law. They just sit there, and she eats. She eats while looking somewhere else and does not say anything.

Wampar language has two further expressions that overlap with 'stingy' and are used less frequently. One, *rai dangi*, has 'to be jealous' as its pivotal meaning (also 'overprotective', 'thrifty').[4] *Rai dangi* is frequently used to talk about people who are possessive or easily become jealous in a relationship; it is used to talk about sexual jealousy or of couples who spend too much time together. It can also be used for money and things: *Yai umu raum dangi en moneng*, 'you are jealous (stingy, thrifty) with money' (*mangalim* something or to be envious, TP). Alternatively, it can be used in the sense that somebody is too firmly attached to things:

---

4   Burbank (2014) discusses translations of 'envy' and 'jealousy' in different languages and sociocultural contexts and the basic inequity aversion and negative social comparison that is expressed in one or more 'emotives'.

> When people see a stingy woman or a stingy man, they say to each other: 'Do not take their things or keep their things because they are a stingy person. Whatever your eyes see, it actually "belongs" to them. If you take it, you will hear from them that it is not right.'

The second expression is *mara gwarog* (literally 'loving eyes', akin to TP *ai gris*), mainly meaning 'to like, to love, to desire': *Ngaeng imu mara gwarog en moneng*, 'the man loves (saves) money'. This phrase can be used like 'stingy' or 'greedy', as well as implying that somebody desires money too much. For men and women being stingy concerning guests, and especially in-laws, is considered particularly negative:

> Another man is like this: When he got some meat, he would not give anything to his parents-in-law or his brother-in-law. When he brings it home, his wife cooks it, and they eat together without giving anything to his in-laws. When they finish eating, he says to his wife: 'Put my leftovers away. When you cook again, I will eat them.'

> If a man with good manners is married to a stingy woman, then when she eats, she just looks down, and does not call other people or the sisters or the brothers of her husband to give them the leftovers. She does not call over other people.

Stingy people avert their eyes (*samasam-eran*) and say nothing when the circumstances raise the matter of sharing. When their attitude is known to fellow Wampar, a stingy person is expected to be ashamed. However, people have different sensitivities and some care more about their reputation than others. Shame or embarrassment are often expressed and given as a reason to stay away from social events for a couple of weeks or months depending on how serious the conflict was. Even if a person is not ashamed, her/his kin might be and expect her or him to maintain a low profile.

Wampar described the opposite of stinginess as 'having good manners, being generous' (*mpe*), which can be said of men and women (*ngaeng* or *afi a mpe*):

> When somebody from another village visits his brother or cousin, but is not given any banana or betel nuts, he might approach the son of his brother or cousin who'll give him something. If his stingy brother or cousin comes to visit him, he still gives them something anyway, so that they eat and chew betel, and drink coconut water. He still takes care of them. This is a man with good manners.

Shaming through gossip (or reputational questioning) was and still is a widespread negative response to non-social behaviour. The most important term for gossip is *yawin* (*tok baksait* TP),[5] 'gossip, defamation, slander, character assassination, untrue tales'. *Edza amu yawin en a gea*, 'I gossip about him'. *Afi imu yawin en eran*, 'The women slandered each other'. *Dzob yawin* can be translated as 'defamatory talk', it covers 'true' and 'false' stories about a person, about which the persons concerned might or might not know about; it is also used in a biblical sense *Oteg a dzob yawin*, 'Do not bear false witness' (The Bible Society of Papua New Guinea 1984). Some Wampar explained that it can have the same meaning as *dzob muam*, 'untruth' or 'untrue tale'. As it says in another text: *Ngaeng Wampar ges etao dzob yawin efa ram a furan ongan*, 'The Wampar see *dzob yawin* as something bad'. At the same time, it is part of endless everyday discussions about the behaviours of others. People accuse each other of *yawin*, and there is *yawin* about people practising too much *yawin*.

Someone ashamed by gossip, bows his/her head or turns away on meeting somebody, or actively avoids others altogether. As Wampar say, an embarrassed person 'hides his/her face'. The proper verb is *meatseran*, 'to be ashamed of, to be embarrassed'; also 'to be shy, to feel embarrassed'. *Edza ameats* then means 'I am ashamed'; the noun is *meats*, 'shame'. There are several specific phrases formed with *meats*, such as possessives, *gea meats*, 'his shame'; *ifu en a meats*, 'he suffers from shame' (or: 'is afraid of shame'); *engap en a meats*, 'he is ruined by shame'; or *erem a meats ari garagab ongan*, 'he gives shame to another man'.[6] *Ngaeng imu ram a furan da emeats egwaro*, 'A man who has done something bad, is ashamed and looks down'. *Ngaeng engop en a meats, esesaran*, 'When a man is ashamed, he turns away'. There are different situations when people feel ashamed: When someone has stolen something, and it becomes generally known,

---

5   There are further terms for gossip: *dzob gangkan* 'insignificant speech' or *dzob inin* 'gossip, tattle'; *mpu-ran* 'to chat, talk glibly, chatter'; *rawedz-eran* 'to talk about many things, chat' (cf. Fischer and Beer 2021).

6   Shame is frequently used to characterise interactions between boys and girls, men and women as well as topics related to sexuality and pregnancy. For instance: *Garafu afi emeats en a garafu maro debareg gentet inin*, 'Girls are ashamed in front of boys and stay away from them'. *Afi uri gea emeats en sun en a ntsigintsigeran*, 'This woman is ashamed of being close to her husband'. *Afi fureran emeats en mpomeran a gab ofo, da esesef a gab inin*, 'Pregnant women are ashamed of walking through the village, therefore they walk only outside the village'. A newly married woman or a woman who has given birth recently is also feeling embarrassed. Women and men among themselves are not ashamed when they are naked, although women should never undress entirely. These feelings of embarrassment have consequences for rules regarding washing in the small streams around the village and the river: washing places for men and women are strictly separated. Although men sometimes pass a washing place of women, they will look away, will not greet, and pretend they have not seen anybody.

not only the thief but also his relatives suffer shame. If people are ashamed of having wronged somebody, they will give food to that person, but only after having stayed away from others for a period. On such an occasion, they shake hands and eat with him and his relatives.

> Imagine a man is ashamed because he shot other people's pig or killed their dog or stole their sweet potatoes, yams or taro. People will abuse him, shame him, and he does not go to them or ask them for anything because he is ashamed. Only later, will he give them food, shakes their hands and eats with them. Then his shame is over.'

Sometimes the shamed person leaves the village for a time:

> If somebody does something terrible and is very ashamed, then he goes away and looks for work with white people. When there are problems with a woman, or he killed someone else's pig or just any pig, then he is humiliated.

In the 1970s, Fischer's interlocutors explained that one 'has shame in the blood' as well as 'in one's head'; it occupies one's thoughts, but is also a bodily sensation. Even if nobody saw who did it, if the news spreads that something has been stolen the thief feels shame, and it takes time until the shame is gone and his 'skin becomes whole again' (orig.: 'Shame is over, and his skin comes back', *Meats empes a gea rene gangkan wasif eama*).

Fischer's texts show that quite specific ideas about 'right' and 'wrong' behaviour were widely discussed and formed an important part of the fabric of sociality. In discussing behaviours and social situations, people simultaneously argued about the facts and about the values they implicate: stinginess, good manners/generosity and greed. These 'thick' emic concepts (Williams 2006 [1985]: 129ff) were used to characterise interpersonal relations based on transfers, all of which are integral to egalitarian social relations. Such concepts, as Bernard Williams emphasised, are evaluative and action-guiding, but also depend upon the facts of people's actions and motivations (ibid.: 140–41):

> Of course, exactly what reason for action is provided, and to whom, depends on the situation, in ways that may well be governed by this and by other ethical concepts, but some general connection with action is clear enough. We may say, summarily, that [thick ethical] concepts are 'action-guiding'. (ibid.: 140)

Being a 'smelly person' provides others with reasons for sanctions of that person's antisocial behaviours through gossip and shaming, which might encourage that person to avoid others or change his or her behaviours.

Today the dependence of individual Wampar on immediate social networks has become weaker than in the 1970s. For most people, it is not difficult to leave the village for some time, they can stay with relatives or friends in distant places or the next town and return to the village after enough time has passed for whatever conflict has cooled down or is even forgotten.[7] With growing economic inequalities, some wealthy people based on their standing and economic influence can also limit to some degree the damage that gossip can do, as discussed below in the case of Matias, a wealthy landowner who has plots in most important places along the highway and owns the land of the market.

Thus, social sanctions have lost something of their effectiveness. However, thick ethical concepts still play an essential role in everyday discourses among Wampar-speakers and affect people's social standing. Their significance is particularly reflected in the socialisation of Wampar children who are continually encouraged to share and not to be stingy. In today's transformed economic environment, these thick ethical notions might still support egalitarian ideals such as sharing with others and being generous in redistributive events, but, at the same time, these events (e.g. children's birthday parties, weddings, funerals) provide opportunities for conspicuous consumption and competition.

Global Entanglements and Their Consequences

I argue that social inequalities tending to develop under increasing capital investment and consumerism in the Markham Valley seem to be inevitable, but their consequences do not unfold as a unilinear process, as some of the economics and political science literature on globalisation and/or 'modernity' presumed (cf. Lewellen 2002): resistance in local contexts is common, and different long-term historical developments reflect local specificities.

---

7   Census data has shown that many Wampar have more than one household in which they stay for some time for economic or other reasons. Several interethnic couples maintain households in the place of origin of each partner and children of such unions can continue to use both affiliations.

Populations in the Markham Valley have interacted with outside forces since the end of the nineteenth century. Early in the 1920s, some Wampar individuals had close contact with the Lutheran mission and to the representatives of colonial administration, which enabled them to access educational resources not available to most others. Access to education, permanent cash income from employment in town, or extra income from investments in raising chickens or in local shops have increased economic inequality over the decades. In 2017, I investigated a random sample of 30 households from my census and found great differences in access to education and health services, as well as in standards of housing (mosquito safe or not), access to clean water and electricity, as well as in the foods consumed. The female head of a migrant's household reported, for example, that she and her children went to bed hungry several nights every week, while others enjoyed modern, well-furnished houses and had rice, noodles and fish or meat, in addition to soft drinks, on a daily basis. Census data from other Wampar villages, collected by Schwoerer and Church shows that the residents of Gabsongkeg were comparatively well-off.

How have Wampar obligations to share and reciprocate, and the negotiations they involve, changed with the introduction of money, consumerism and growing inequalities in the context of anticipated wealth from mining? I will describe in more detail the case of an in-married former medically trained woman who has been active in PNG's Women's Micro Bank Limited (WMBL), as well as in the development of children's birthday and fundraising parties as new communal events of distribution. I discuss the consequences of (perceived) violations of expected behaviour in transfers and the challenges to, and persistence of, thick ethical concepts under conditions of growing social and economic inequalities.

Disputed Behaviours: A Case Study

Gertrud is from a province in PNG, which, like the Markham Valley, is an early contacted and modernised coastal area with relatively high prestige in national, ethnic hierarchies. She met her future husband Topom-Matias in 1988 during a basketball tournament in one of PNG's major mining sites, where he worked as an engineer. The couple moved into her home because housing in the mining compound was for men only. A year later, when Gertrud was expecting their first child, the couple separated and went back to their respective relatives. Some months later, after Gertrud had given birth to their son, she followed Topom-Matias to Gabsongkeg. In 2014, she worked in one of Lae's hospitals while her husband had a series of contracts at various mines. The couple had two more children, built a house and settled in Gabsongkeg. Gertrud and Topom-Matias are exemplary of an emerging PNG middle class and their aspirations (cf. Gewertz and Errington 1999). For extended periods, they lived in PNG's mining sites, where Topom-Matias had found work. Gertrud began to import and sell garden products from the Markham Valley to families of workers living at the sites. Topom-Matias used to say that his children have 'the best of both worlds', for they can inherit land in Gabsongkeg and in Gertrud's village of origin, where land ideally stays in the matriline.

Gertrud joined her husband on and off for several months at the mining sites, alternating with extended periods with their children in Gabsongkeg. Topom-Matias died while working at a mine. Since his death, Gertrud has remained with her three children in Gabsongkeg, although in 2018 only her 16-year-old son remained at home. He attended a new private primary school near the Highlands Highway run by Seventh-day Adventists because his mother believed it is superior to the public school in Gabsongkeg. Both her older children moved to Lae after their marriages.

In 1995, Harry, one of Gertrud's brothers, who had also married a Wampar woman and was living in Gabsongkeg, became ill during a visit to their village of origin. After his return to Gabsongkeg, his health improved, but he died suddenly one night in 2007. Some Wampar attributed both Topom-Matias's and Harry's deaths to sorcery. This reflects a more general Wampar view that powerful sorcery is prevalent in particular ethnic groups, which is itself supported by migrants' sometimes tense relations with relatives back home, as is common in PNG. Gertrud herself said it is generally dangerous for migrants to visit their relatives back home because of jealousy and the availability of sorcery. Some of her neighbours were more specific and said that Gertrud's uncle used sorcery on those who declined his requests for financial support. Sharing income and/or consumer goods with relatives back home is an expectation that migrants cannot always meet; and they are in a position to discount or ignore it if they want. Several migrants anticipated jealousy, anger and the danger of sorcery for that reason when they plan visits home. Gertrud herself declared that she would not take her children to her village of origin, out of fear for their health, and has never done so. Some of her relatives have, though, gone for visits and to attend funerals in Gabsongkeg.

Discussions among other Wampar about Gertrud's behaviour during visits of her relatives included suggestions that she did not look after them properly, so that her Wampar affines had to do it. A woman told me she thought that Gertrud had not visited her village of origin because she is no longer welcome, adding that her relatives knew of—and here she used the English phrase—Gertrud's 'personality problem', her self-centredness and want of generosity. Affinal relatives also criticised her for beating her children and refusing to treat them to little snacks or gifts from town. In short, she was said to be stingy, reluctant to help others and pursued her own financial and business interests and plans.

What else was behind Gertrud's 'personality problem'? As an educated woman, she often emphasised that Wampar are unwilling to invest the time and effort required to improve their standard of living. Still, she said, she enjoys life in the village. Gertrud did not want to go back to work in Lae, despite her professional qualifications, because she did not want to live in town or to commute; she felt that village life would be better for her children (and grandchildren). Instead, she turned to the small business opportunities that living in Gabsongkeg offers: raising and selling chickens, planting cocoa or melons and, lately, establishing a small, permanent shack at the Nadzab market from which she sells food and drinks. Most mornings, at an early hour, she could be seen waiting on the little forest road leading from Gabsongkeg to the highway for transport to take her and her wares to the market. Many Wampar criticised her rigid commitment to these activities, as was shown one morning when Gertrud waited for transport, at dawn, near the compound of a close relative of her in-laws who had died during the night. Mourning was just about to start. Traditionally, the market would be entirely closed at least for a day after mourning ceremonies began; this is no longer the case, but those close to the deceased are still expected to join in when mourning starts (cf. Beer and Church 2019).[8] A senior, influential man criticised Gertrud in public for her attempt to continue 'business as usual'. Another added *Em i stap bilong en yet*, which means 'she only thinks of herself'. Female relatives

---

8   Several changes contributed to these new marketing practices: the market is no longer used only by Gabsongkeg vendors, the owner of the land does not care much for the church community in Gabsongkeg and as a result the appointed *maket komiti* enforced market closures only rarely and a growing number of people ignore them. Still, many Wampar families expect that relatives of a deceased person do not go to the market to sell their products but take part in the preparations for the funeral.

discussed her failure to contribute to meals in her in-law's household, where she frequently ate in the evenings. Portrayed as a classic free rider, her reputation as *afi muteran* was widespread.[9]

I heard rumours that Gertrud privately lent money to fellow Gabsongkeg people at high rates of interest, but I could not confirm this. She was, though, an active member and treasurer of the Gabsongkeg group of the Women's Micro Bank Limited (WMBL), which, since 2014, has been licensed and regulated by the Bank of PNG.[10] Gertrud visited its crowded town office regularly and was head of the local members' 'group' in the village, but when I asked her about WMBL's interest rates, she could not tell me. A Wampar who had worked for many years in a bank in Port Moresby explained that the WMBL did not pay any interest on savings.

Gertrud was very excited that the bank would release shares in its own finance company and that they would be available for everybody soon.[11] Nobody, though, could explain the advantages of the shares, or what shareholders might get in return in the future. The same applied to shares people could buy of the landowner business group that will run the PNG Biomass plantations (see Schwoerer, this volume). Gurum, an elderly Wampar woman who had a WMBL savings book since 2014, for example, used 300 of her 551 kina on her account to buy shares, but had only an entry in her passbook to prove it. Since 2014, Gurum has paid 119 kina in fees to WMBL, and paid 300 for shares, but she was unsure how to access her remaining funds if the need arose. These examples do not reflect on Gertrud, but the inscrutability of banking practices help explain some of the villagers' suspicion that Gertrud acts in questionable ways.

9   I also came into conflict with her because she broke a promise to help an older, nearly blind female relative keep an appointment with an eye doctor in Lae, on a day when I could not take her. I and others expected her to do what she promised, especially as she received medical education and understood the importance of continuity of treatment.
10   The chairperson of this group was the wealthiest woman in Gabsongkeg, whom I will introduce in the next section.
11   See the WMBL brochure (www.womenmicrobank.com/wp-content/uploads/2017/07/shares_brochure.pdf), which said that they were only available till the end of November 2017. In October 2019 the WMBL received a loan of PGK830,000 (USD243,785) from the UN Capital Development Fund (UNCDF) (www.uncdf.org/article/5077/womens-microbank-in-papua-new-guinea-receives-loan-from-uncdf).

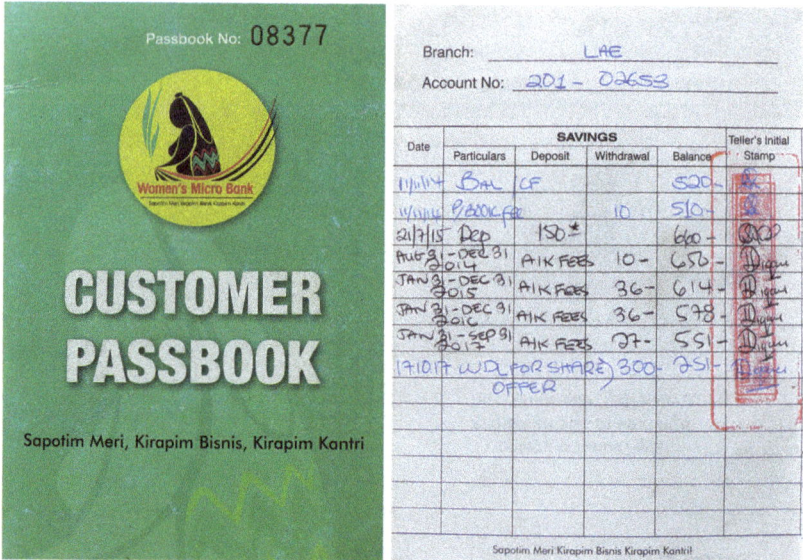

Figure 5.1 Gurum's savings book for the WMBL.
Source: Photos by Bettina Beer, 2017.

Financialisation in Gabsongkeg included not only the WMBL, but also the opening of a branch of the Bank of South Pacific in Anna's store, who is one of the wealthiest persons in Gabsongkeg; founding of and activities by the Gabsongkeg Resource Owners Association (GROA); and emerging business initiatives such as the Gabsongkeg Development Foundation (GDF). Some fast money scams also operated in Gabsongkeg, such as the initiative of an individual who collected money from fellow villagers to prosecute Japan to pay compensation for damage done during the Second World War. Many Wampar cannot see a difference between WMBL and other initiatives promising financial returns. John Cox (2018: 51) describes the similarity between pyramid selling, fast money scams and microfinance schemes: they 'draw on similar visions of prosperity and present themselves as a reliable route out of poverty'. Caroline Schuster (2015) analysed the entanglement of a Ponzi scheme elite and the microfinance non-governmental organisation (NGO) she studied in Paraguay. Authors working in South America, South Africa or Melanesia on schemes, scams and gambling show that distinctions between legitimate and illegitimate financial spheres are notoriously difficult to draw and to sustain. These (fast) money schemes thrive in the discursive space of development, poverty and stark wealth differentials. It is not surprising, in the context of the increasing presence of companies, land sales and thoughts about the ever-larger sums of money flowing through Morobe Province, that desires for monetary wealth find a fertile ground to grow.

Gertrud told me that she knew people gossiped about her. She explained it in terms of the contrast between her activities and those of her peers: her involvement with diverse projects based outside the village make it often impossible for her to join communal activities. Gertrud also claimed that many Wampar do not understand how, through their own efforts, they could improve their economic situation, complaining that only a few women had followed her recommendation and opened accounts with the WMBL. Gertrud herself noticed that her perceived indifference to expectations of reciprocity in the broader social field affected her reputation, but she regarded fellow villagers as too backward to understand her work. In the postcolonial context of large-scale capital projects, when it comes to control of family budgets, to investments and to land access, use and sales, gender relations become conspicuous in the context of growing inequalities (described in detail for Wampar of the Gabsongkeg area in Beer 2018).

Money scams such as U-Vistract (Cox 2018) or the above-mentioned Japanese war compensation scam, microloans and other financing instruments promising 'development' have been mushrooming in PNG over the last 20 years. At the same time, ways of life have diversified as a result of differential involvement with capitalist economies. This has led to differing positions in negotiations of transfers in social relations: a shop owner cannot give her goods to relatives for free, and people mobilise their networks for new economic projects. Complaints about Gertrud's behaviour are not so different from the examples given to Hans Fischer: she does not help her in-laws or visitors as she should and takes from others more than she gives. Problems of free-riding and stinginess can be found in any social world, but her persistent attempts to involve others in new schemes beyond their understanding has no precedent in Fischer's work of the 1970s to 1990s. Today, the amount of money available and growing need for it makes the advantages and costs of free-riding greater and more apparent. The work with NGOs and different kinds of government-funded projects reflects new social inequalities in education and access to information and institutions on a regional and national level. Although Gertrud is not the only woman working with the WMBL or comparable financial institutions, her reputation ('personality problem') leads to more suspicion and gossip than in cases of other women working with the bank.

Children's Birthdays and Fundraising Parties

A story of social change from traditional community spirit to individualistic behaviour and the 'decline of traditional values', as expressed by some Wampar, is an integral aspect of contemporary ethnography of social change. Several types of communal feasts (e.g. bridewealth, funerals) are still held, and coexist with newly established social events (modern weddings, Lutheran church events like *sanisim basket*[12] or fundraising parties), which emphasise exchange, mutual help and sharing. In a context of consumerism and the negotiation of relations, these new events demonstrate that people remain concerned with maintaining good relations with neighbours, friends and kin, despite the accumulation of financial and cultural capital that tends to entrench inequalities. They open possibilities for sharing as well as for conspicuous, competitive consumption. Thus, sharing and redistribution should not be understood as antithetical to inequality. On the contrary, events in which they feature contribute to the growing significance of class distinctions in PNG. However, they can provide temporary relief as they offer food for people in need, such as disabled, non-married or older people whose gardens provide only the minimum food security. Simultaneously, they are underlining the growing differences as a social, acknowledged fact.

Since roughly 2000, some families have begun to celebrate their children's birthdays by throwing a party; not every year, or for every child, but when well-off parents (in agreement with the extended family) decide it is time to do something special and invite others to participate.[13] Sometimes, a family sends out a formal invitation card. Other tokens of modernity and middle-class affluence, such as personalised birthday cakes or video documentation of the event, also feature.

---

12   These are events when members of two congregations or church sub-groups meet to begin or end a project; for example, to formally open a new church building, have a workshop or visit each other. Both sides take food and net bags or baskets to exchange. The church has its novel history of emphasising sharing and other values, especially given that Christianity has a complicated history with authority and wealth. Pentecostalism, new evangelical born-again churches add further dimensions (Jorgensen 2005; Cox and Macintyre 2014).

13   The decision to celebrate a birthday is similar to the decision to present bridewealth, although the latter involves two families coming to an agreement.

Figure 5.2 Formal birthday party invitation card.

The reverse of the card reads: 'You are cordially invited to attend NAME's 10th Birthday Party to be held at NAME's place, Gabsongkeg Village [Nadzab] on SATURDAY 02 March 2002 Time 2.00 PM rsvp NAME and phone number.'

Source: Photo by Bettina Beer, 2002.

Elaborate birthday parties are reported to be part of new middle-class repertoires in such different contexts as PNG, Namibia (Pauli 2018) and the US (Clarke 2007). The emphasis on parent–child relationships is central to notions of the family, and are taken to be closely connected to the advent of consumerism.[14] Among the Wampar, birthday parties do not (yet) include personalised gift-giving by the guests outside the nuclear family; they centre on the distribution and consumption of mostly modern drinks and foods (including personalised birthday cakes, bought in town), although hosts offer some traditional foods as well.[15] Invitation cards (Figure 5.2) are also part of the new middle-class repertoire. Yet poor relatives and neighbours are not excluded from such feasts, even if the order in which they are served indicates their relative standing.

---

14 A 'modern'—very exceptional—wedding with specific wedding attire of the couple, a cake and a feast with much food and many guests has been celebrated (and photo documented) in Gabsongkeg too (cf. for social positioning, class and weddings in Namibia Pauli 2018, 2019).

15 In Dzifasing some of the guests give birthday gifts.

However, birthday parties and modern, elaborate mourning ceremonies are not straightforward levelling mechanisms vis-à-vis emerging social inequalities; they are also part of the growing importance of consumerism. These patterns of conspicuous consumption also contribute to the growing emphasis among the Wampar of the nuclear family as a central unit in social life. Related ideas and practices tend to limit moral obligations to a smaller unit than the patriline and *sagaseg*-centred networks, which were the main point of reference in many realms of everyday life. On the other hand, egocentric networks are extended by these new social events, as neighbours, distant and in-married relatives, friends, co-workers or temporary visitors are part of the celebration. They are invited and help with the preparation of food, while the communal meal is the most crucial part of the event (which might last a whole day).

Fundraising parties are another new type of event, the point of which is to help individual families raise money for school fees. Education is seen by most Wampar as crucial to upward mobility and has been proven to be so since the 1920s, when missionaries established the first bible and later elementary schools in the area. The payment of school fees (and often of parents' additional contributions, when teachers do not receive pay, buildings have to be repaired, or toilets built) have been a constant concern of people in PNG.[16] Higher secondary and tertiary education, which require living away from home, involve particularly high costs for parents who have restricted access to wage labour and money-making opportunities. Since around 2007, when the betelnut economy collapsed, households with children of school age have faced significant financial difficulties in educating their children.

In order to help meet education costs, parents and their close kin have begun organising big meals as 'fundraising parties'. Guests make a monetary contribution, although some might contribute garden products. The foods usually consist of banana and garden products, with the addition of rice and some meat such as sausages, chicken or pork, if funds allow; the form and spirit of the get-together are very similar to

---

16   In 2012, Peter O'Neill introduced a Tuition Fee Free (TFF) policy, which has never really worked in many regions (cf. edu.pngfacts.com/education-news/tuition-fees-two-years-late-for-school). Delayed payments of teachers, neglect of school buildings and sanitation made parents' work and financial contributions unavoidable even during TFF times. Since 2018, tuition fees have been reintroduced in several places.

other organised collective meals. The event involves the sharing of work, the provision of garden produce and, most importantly, the collection of money for individual projects.

'Fundraising parties' are not evaluated in the way Gertrud's activities are: many Wampar see them as a legitimate way to share the work and burden of the money economy. Fundraising parties not only show that Wampar see the life chances of future generations as depending on cash income and education, they also show that old and newly created opportunities are used for conspicuous consumption, extending networks beyond the village, which are also seen as a part of contemporary social life. Today's collective meals at bridewealth celebrations, weddings, birthday parties or funerals are social events involving between 50 and a few hundred people. They are much bigger than birthdays or fundraising parties and have grown with the money economy. Funerals can last for up to a week, long enough to allow migrant Wampar to return from other provinces to the village. During the first day and night food is served to all visitors. Related families send garden products, pigs and cash in advance and the women who will help to prepare the food likewise gather at the household where the funeral is to be celebrated many hours before it begins. The hosting of church gatherings is another significant communal event, as when, for example, the Evangelical Lutheran Church (ELC) hosted the District Youth Conference in Dzifasing in 2017: the whole village contributed heaps of bananas, and the more prosperous villagers donated large amounts of cash or cows and pigs, which were consumed by Dzifasing villagers and the attendees of the church event.

Collective meals were mentioned in Fischer's texts on violations of reciprocity in transfers of food or money, as means to restore 'moral equivalence', to end a period of shame, or mark the reconciliation of families. At local village courts, magistrates' decisions regarding conflicts nearly always include the organisation of a joint meal in addition to the payment of a fine. In the current situation, they have become another opportunity display wealth and underline social differentiations that have not existed before. All these examples show that collective events have grown into important occasions to display social status, defend a position in social hierarchies or to aim at upward mobility in a context of growing social inequalities.

The Local Elite, New Middle Classes and Discourses on Not-Sharing

As differences in wealth among Wampar families have become more apparent, feelings of relative deprivation have spread too. As the gap between levels of desire for new consumer goods and the means to get them has steadily widened for most but not all Wampar, discussions of values and the behaviours that they should motivate have become more frequent; egoistic behaviour has become more often subject to gossip, and general reflections about social change are part of everyday conversations. Often, they are triggered by imaginations of what the future will bring when 'mining comes'. Social change and the diversification of values are issues repeatedly raised by Wampar themselves, and generalisations about their social behaviour are more difficult to make than in former times. Wampar reflections on specific changes of behaviour and values provide claims about and evaluations of shared (and non-shared) desires and beliefs, and the grounds of social interaction (cf. Beer and Bender 2015).

Discourses of envy, jealousy and stinginess in the context of social change are not only crucial to relatively impoverished Wampar, but also to the wealthiest Wampar families, who stress the scale and importance of the projects they sponsor as expressions of their concern for 'the community' and their wish to share. They are still embedded in one social field with people who have less and must handle jealousy and inequality in everyday life.

When I surveyed Gabsongkeg Wampar about wealthy families, all named the same four individuals, although the order in which they were listed varied slightly.[17] Everybody agreed that there were four 'wealthy' residents: Matias, Anna, Seref and Yadzu—three men and one woman. I knew all of them well from earlier fieldwork, but during my current research, it was difficult for me to make appointments for lengthy discussions with them. Only after repeated attempts, and enlisting the moral support of their

---

17 We conducted a household survey in 30 households in four of the Wampar villages, which in Gabsongkeg included one of the wealthiest persons. Our survey shows that differences within Gabsongkeg between poor and wealthy households are substantial and that the agreement of Wampar from Gabsongkeg on the wealthiest four matches the results of my survey interviews.

kin, could I get three of these long-standing acquaintances to find time to talk to me; Matias, one of the four, in fact, never granted my request for a meeting.[18]

Matias is one of the wealthiest landowners: his plots are large and in a very advantageous location; moreover he owns a transport company and receives rent from the stallholders at the busy Nadzab market. Furthermore, he has sold and leased a lot of prime land along the Highlands Highway (cf. similar arrangements in peri-urban Melanesia; Mcdonnell et al. 2017).

Matias consistently avoided direct contact with me: his relatives said he was not home or was ill when I called at his house in person, and when I called his phone, he said he could not speak because he had to deal with one urgent family problem or another. Matias's household is located outside the main village, near the Highlands Highway, and he rarely attends public meetings, Sunday service or other social events. He has created an image of a person outside the usual moral networks; only ever in the company of a few close relatives (mostly sons), he seemed separate from the rest of the village. When he is seen in public, he is usually drunk. In 2017/18 he was involved in an ongoing court case resulting from accusations of incest, which were, no doubt, the family difficulties he mentioned. Gossip and rumours seemed not to bother him or his sons very much, and I sometimes had the impression he even fostered them.

I managed to conduct formal interviews with two of those ranked among the wealthiest in my surveys—Anna, a woman in her early 40s, and Seref, a man in his late 50s—only after repeated requests that they make time in their busy schedules for me. Both interviews were striking for the emphasis these entrepreneurs placed on their Christian faith and their commitment to the advance of the 'Gabsongkeg community'. Their strongly felt need to 'give back' was a topic both raised several times during the interviews. During my stay in Gabsongkeg, it became clear that they did not just talk about 'giving back to the community'; in fact, they financed and organised several projects that solved current village problems. When, for example, the narrow gravel road covering the 3 kilometres between Gabsongkeg and the main highway needed repair, Anna used her trucks and employees to transport stones and fill the worst holes; she established a well for safe drinking water, which was publicly accessible and planned

---

18   In 20 years of fieldwork, this was the first time anybody refused outright to support my work by talking to me.

more; she held the position of treasurer in the Gabsongkeg Resource Owner Association (GROA); was chairperson of WMBL; and gave her time to other demanding communal tasks. Likewise, Seref, her male counterpart, had purchased a large passenger-truck for the ELC Wampar Seket,[19] which he drove himself when it was required. He also worked for GROA and took part in many public social events, providing food, firewood, transport and whatever was needed to ensure their success. Both, of course, also benefited from their commitments: Anna used the public road to transport goods to and from her stores in the village and at the airport. Seref and his wife took part in a church-organised tour of Israel. Both affirmed their commitment to Christian morality in general and to the long-established (and still influential) Lutheran Church in the village. Anna, the businesswoman, for example, refuses to allow alcohol to be sold in her three stores.

Although Yadzu is the fourth wealthiest person named in my surveys, he is less visible in community activities and associations than those discussed above. He runs the most successful cocoa fermenter in Gabsongkeg and owns its most lucrative bottle shops. I talked only briefly to him as he is a very busy man and spent much time in town; his wife and several relatives working for him also granted me interviews. This man has 'adopted' several unrelated Wampar and non-Wampar—he looks after them and their families, sharing the fruits of his land and income with them. They all emphasised that his fermenter helped Gabsongkeg cocoa growers to get better prices; his entrepreneurial activities and the network of relatives and workers it embraces was presented as a successful business model.

Interviews showed that several newly rich were preoccupied with 'giving back to the community' and that they were aware of the criticism that violations of moral equivalence provoked, which is not to suggest that helping was instrumentally motivated. Resulting attempts to improve problems in the village (water, employment for relatives, sharing of food at larger events) cannot really mask the consequences of social inequalities and help disadvantaged families, although the gap between haves and have-nots widens remarkably, as our survey data of 30 Wampar households in Gabsongkeg shows. Keir Martin (2013) described land politics, the emergence of family-based interests and the use of the term 'big shots' as an evaluative term for the new rich (and which contrasts

---

19   Other Wampar said the local Member of Parliament (MP) had purchased it.

with the 'traditional big men') after the volcano eruption in Matupit on the Gazelle Peninsula. He suggests that the moral dilemmas based in demands for reciprocity under transformed economic circumstances have not been acknowledged sufficiently in the literature on Melanesia (ibid.: 25). I have never heard 'big shot' in Gabsongkeg, although the critique of corrupt politicians and regional elites has a lot in common with discourses in Matupit. Bruce Knauft (2007: 68) argues that among Tangu (as described in Kenelm Burridge's ethnography) the moral system in place is both intensified and challenged as a dimension of a broader historical dialectic.

Conclusion

In my contribution to this volume, I describe changes in transfers of food, money and consumer goods as they are both related to and implicated in growing social inequalities that are developing under increasing capital investment and consumerism among Wampar and immigrants in a suburbanising location near the city of Lae in PNG. Violations of social expectations concerning reciprocity and the sanctioning of them are not new among Wampar-speakers, as the early texts recorded by Hans Fischer show. Today, the amount of money available and the growing need for it make the advantages and costs of free-riding greater and more apparent.

Life chances (education, medical treatment, transport) depend increasingly on cash, and its unequal distribution impacts on social inequalities in future generations. In the current situation, new dimensions of social inequalities emerge: wealthy people, whose income and livelihoods do not depend as much on the support of others to clear a garden and live off its products as before, have more resources to extend their networks outside the context of the village. Thus, they are less dependent on social support and can be more ruthless in advantaging themselves even more: the rich can get richer without really trying, while the poor go backwards relative to the wealth (which itself generates ill-feeling). They can also use the opportunities of traditional social events such as funerals, shared meals after conflicts, or the handing over of a bride price, as well as new birthday parties, weddings or sports events, to not only display conspicuous consumption but to confirm their 'middle-class' position in today's social hierarchies. Their motivations and desires to do so result from the described changes and are not independent of the context in which they manifest themselves.

Wampar and widespread national discourses in PNG on 'elites, new middle classes and the grassroots' or 'tradition (*kastom* TP) and modernity' show different ways people try to position themselves and see others in a context of rapid economic and social change. Bruce Knauft (2007) uses Burridge's notion of 'moral equivalence' in his comparison of social change among Tangu and Gebusi to refer to mechanisms to restore and maintain balance in social relations. Negotiations around the equivalence of persons and sharing were and are still the basis for sociality in Wampar networks, although the significance of cash income and the monetary economy pose new challenges, as described above. Many Wampar are themselves aware of the transformation of motivations that has accompanied their increased integration into the nation state and a market economy: nowadays some actively try to avoid or curtail expectations that were integral to their social relations and networks only 20 years ago.[20] Gertrud's attempts to establish a small business and the gains she sought from the financial market are read in these terms, as is the behaviour of Matias, the wealthy landowner who refused to discuss his businesses with me, who does not follow basic communal rules and in many respects positions himself beyond village sociality. By contrast, other members of the local elite are concerned with the state of the community and express the need 'to give something back' to the community.

Gossip, blaming and shaming were effective means to maintain social cohesion and still bother some Wampar today; complaints of selfishness and threats of social sanctions are present. People often reacted with withdrawal followed by reintegration. The degree to which these sanctions have teeth has changed as Wampar are no longer as dependent on their immediate social network as before. After disputes, from small conflicts to severe crimes, Wampar often leave the village and stay with relatives or friends in town or another province. Census data collected over the last decades strongly supports this observation. Some families—due to the increasing number of interethnic marriages—even maintain households and close social networks in two places a long way apart. Or, as the example of one of the wealthiest landowners shows, with enough resources and

---

20   This was especially a vital lesson to learn for shop owners in the villages who at first wanted not to give all their goods away to kin, but be able to start a small *bisnis* (cf. Curry 2005).

support from people who depend on them, one can even stay and cut off relations outside the closest network. Today anti-hierarchical strategies are failing to contain rising inequality. In the past, being socially sanctioned and isolating was not merely shameful, it was also potentially deleterious. With the amount of money flowing around, a wealthy person does not need as many social connections (or can be more selective about what they are). Pairing changing moral discourse with the changing material necessities adds a new dimension of political economy to the discussions on ethics under capitalism.

The ethnography shows how different segments of local social fields can or cannot engage the encompassing global processes to different degrees and in different ways, depending on a host of social factors. The upshot of such initial differentiating processes is frequently the production of significant social and economic demarcations, which itself is crucial to the generation of more entrenched social contrasts in the medium and longer term.

I suggest that the social inequalities developing under increasing capital investment linked to international markets and spreading global consumerism in the Markham Valley are one reason among others for the changes in social relations described. Emerging social hierarchies are also based on colonial and mission history, as well as the geography of economic differences related to the proximity to the Highlands Highway (Beer and Church 2019) and Lae city, with their markets and locally circulating capital. The village of Gabsongkeg has a prominent position related to all these conditions: it was closest to the former mission station, to the Lae city airport and one of the biggest highway markets on the way to the highlands. Growing desires for consumer goods, which are less frequently shared than, for example, garden products, are implicated in today's perceived violations of reciprocity, theft and fraud, as well as in hasty investments in various 'fast money' schemes, or the establishment of risky business ventures. They also lead to new competitive social events, such as children's birthday parties, weddings and fundraising parties for school fees, which have become part of the aspirations of PNG's middle classes. Furthermore, increasing cash flows and consumerism increase the efflorescence of traditional distributive events like funerals, the hosting of church conferences, or the organisation of and participation in sports events. As feelings of relative deprivation have spread, and the gap between the desire for goods and the means to get them has steadily widened, discussions of values and the behaviours that they should motivate have become more frequent. These discourses are critical among relatively impoverished Wampar, as well as among the wealthiest, who seek to emphasise the scale of the projects they finance and the concern for the community that motivates them.

In 2007 John Barker wrote that the primary concern in the Melanesian ethnography of morality is with the interface between 'indigenous village life and the ethical orientations associated with "modernity"' (2007: 1). Our ethnography of the emergence of social inequalities shows that notions of '*kastom*' and 'modernity', the 'local' and 'global', have become part of Wampar concerns, but are too vague to analyse mechanisms of transformations of social fields. More than 10 years after the publication of Barker's volume, questions of ethical orientations of migrants, local middle classes, national elites and the growing social inequalities in the political economy associated with a region strongly influenced by mining revenues have become even more prominent.

References

Antweiler, C., 2016. *Our Common Denominator: Human Universals Revisited.* New York; Oxford: Berghahn Books. doi.org/10.2307/j.ctvpj7h57

Barker, J., 2007. 'Introduction: The Anthropological Study of Morality in Melanesia.' In J. Barker (ed.), *The Anthropology of Morality in Melanesia and Beyond.* Farnham, Burlington: Ashgate.

Beer, B., 2015. 'Cross-Sex Siblingship and Marriage: Transformations of Kinship Relations among the Wampar, Papua New Guinea.' *Anthropologica* 57: 211–224.

———, 2018. 'Gender and Inequality in a Postcolonial Context of Large-Scale Capitalist Projects in the Markham Valley, Papua New Guinea.' *The Australian Journal of Anthropology* 29(3): 348–64. doi.org/10.1111/taja.12298

Beer, B. and A. Bender, 2015. 'Causal Inferences About Others' Behavior Among the Wampar, Papua New Guinea—and Why They Are Hard to Elicit.' *Frontiers in Psychology* 6: 128. doi.org/10.3389/fpsyg.2015.00128

Beer, B. and W. Church, 2019. 'Roads to Inequality: Infrastructure and Historically Grown Regional Differences in the Markham Valley, Papua New Guinea.' *Oceania* 89(1): 2–19. doi.org/10.1002/ocea.5210

Burbank, V., 2014. 'Envy and Egalitarianism in Aboriginal Australia: An Integrative Approach.' *The Australian Journal of Anthropology* 25: 1–25. doi.org/10.1111/taja.12068

Burridge, K., 1969. *Tangu Traditions. A Study of the Way of Life, Mythology, and Developing Experience of a New Guinea People.* Oxford: Clarendon Press.

Busse, M. and T.L.M. Sharp (eds), 2019. 'Market Places and Morality in Papua New Guinea.' *Oceania Special Issue* 89(2): 126–153. doi.org/10.1002/ocea.5218

Clarke, A.J., 2007. 'Making Sameness: Mothering, Commerce and the Culture of Children's Birthday Parties.' In E. Casey and L. Martens (eds), *Gender and Consumption: Domestic Cultures and the Commercialisation of Everyday Life.* Aldershot: Ashgate.

Cox, J., 2018. *Fast Money Schemes: Hope and Deception in Papua New Guinea.* Bloomington: Indiana University Press. doi.org/10.2307/j.ctv6mtfjm

Cox, J. and M. Macintyre, 2014. 'Christian Marriage, Money Scams, and Melanesian Social Imaginaries.' *Oceania* 84(2):138–157. doi.org/10.1002/ocea.5048

Curry, G.N., 2005. 'Doing "Business" in Papua New Guinea: The Social Embeddedness of Small Business Enterprises.' *Journal of Small Business & Entrepreneurship* 18(2): 231–246. doi.org/10.1080/08276331.2005.10593343

Fischer, H., 1975. *Gabsongkeg 71. Verwandtschaft, Siedlung und Landbesitz in einem Dorf in Neuguinea* [Gabsongkeg 71. Kinship, Settlement and Land Rights in a Village in Papua New Guinea]. München: Kommissionsverlag Klaus Renner (Hamburger Reihe zur Kultur- und Sprachwissenschaft, Band 10).

———, 1992. *Weisse und Wilde. Erste Kontakte und Anfänge der Mission* [White Men and Wild People. First Contacts and the Beginnings of Evangelisation]. Berlin: Reimer (Materialien zur Kultur der Wampar, Papua New Guinea, 1).

Fischer, H. and B. Beer, 2021. *Wampar–English Dictionary. With an English–Wampar Finder List.* Canberra: ANU Press. doi.org/10.22459/WED.2021

Gewertz, D. and F. Errington, 1999. *Emerging Class in Papua New Guinea: The Telling of a Difference.* Cambridge: Cambridge University Press. doi.org/10.1017/CBO9780511606120

Jorgensen, D., 2005. 'Third Wave Evangelism and the Politics of the Global in Papua New Guinea: Spiritual Warfare and the Recreation of Place in Telefolmin.' *Oceania* 75(4): 444–461. doi.org/10.1002/j.1834-4461.2005.tb02902.x

Keane, W., 2016. *Ethical Life: Its Natural and Social Histories.* Princeton: Princeton University Press. doi.org/10.1515/9781400873593

Knauft, B.M., 2007. 'Moral Exchange and Exchanging Morals: Alternative Paths of Cultural Change in Papua New Guinea.' In J. Barker (ed.), *The Anthropology of Morality in Melanesia and Beyond.* Farnham, Burlington: Ashgate.

Laidlaw, J., 2014. *The Subject of Virtue: An Anthropology of Ethics and Freedom.* Cambridge: Cambridge University Press. doi.org/10.1017/CBO9781139 236232

Lewellen, T.C., 2002. *The Anthropology of Globalization: Cultural Anthropology Enters the 21st Century.* Westport: Bergin and Garvey.

Mahmood, S., 2012. 'Ethics and Piety.' In D. Fassin (ed.), *A Companion to Moral Anthropology.* Hoboken, New Jersey: Wiley. doi.org/10.1002/9781118290620. ch13

Martin, K., 2013. *The Death of the Big Men and the Rise of the Big Shots: Custom and Conflict in East New Britain.* New York: Berghahn Books.

Mattingly, C., 2014. *Moral Laboratories: Family Peril and the Struggle for a Good Life.* Berkeley: University of California Press.

Mattingly, C. and J.C. Throop, 2018. 'The Anthropology of Ethics and Morality.' *Annual Review of Anthropology* 47:475–492. doi.org/10.1146/annurev-anthro-102317-050129

Mcdonnell, S., M.G. Allen and C. Filer (eds), 2017. *Kastom, Property and Ideology: Land Transformations in Melanesia.* Canberra: ANU Press. doi.org/10.22459/KPI.03.2017

Pauli, J., 2018. 'Pathways into the Middle: Rites of Passage and Emerging Middle Classes in Namibia.' In L. Kroeker, D. O'Kane and T. Haeberlein (eds), *Middle Classes in Africa: Changing Lives and Conceptual Challenges.* Bayreuth: Palgrave Macmillan. doi.org/10.1007/978-3-319-62148-7_11

———, 2019. *The Decline of Marriage in Namibia: Kinship and Social Class in a Rural Community.* Bielefeld: Transcript. doi.org/10.1515/9783839443033

Read, K.E., 1955. 'Morality and the Concept of Person Among the Gahuku-Gama.' *Oceania* 25(4): 233–282. doi.org/10.1002/j.1834-4461.1955.tb00651.x

Robbins, J., 2013. 'Beyond the Suffering Subject: Toward an Anthropology of the Good.' *Journal of the Royal Anthropological Institute* 19: 447–462. doi.org/10.1111/1467-9655.12044

Schnegg, M., 2015. 'Reciprocity on Demand: Sharing and Exchanging Food in Northwestern Namibia.' *Human Nature* 26: 313–330. doi.org/10.1007/s12110-015-9236-5

Schuster, C., 2015. *Social Collateral: Women and Microfinance in Paraguay's Smuggling Economy*. Berkeley: University of California Press. doi.org/10.1525/9780520962200

The Bible Society of Papua New Guinea, 1984. *Jaer Garaweran Jesus Kristus agea fenefon ngarobingin akani 1984: The New Testament in WAMPAR*. Port Moresby: The Bible Society of Papua New Guinea.

Widlok, T., 2013. 'Sharing.' *HAU: Journal of Ethnographic Theory* 3(2): 11–31. doi.org/10.14318/hau3.2.003

Williams, B. 2006 [1985]. *Ethics and the Limits of Philosophy*. London; New York: Routledge. doi.org/10.4324/9780203969847

Woodburn, J., 1998. '"Sharing is Not a Form of Exchange": An Analysis of Property-Sharing in Immediate-Return Hunter-Gatherer Societies.' In C.M. Hann (ed.), *Property Relations*. Cambridge: Cambridge University Press.

6

Absent Development as Cultural Economy: Resource Extraction and Enchained Inequity in Papua New Guinea

Bruce Knauft

Coin of the Realm

During the past four decades, mining and oil/gas developments have increasingly become the centrepiece, the Holy Grail, of economic and social development in Papua New Guinea (PNG). This is highly evident in national-level discourse and in local desires for mega-development. One may take by example an eight-page full-colour PNG advertising spread in *The Wall Street Journal* (*WSJ*), which came to news-stands in the US in November 2018 (Eye on PNG 2018). Based on published advertising rates for the *WSJ*, the newspaper insert cost about USD2 million, or about 6.5 million kina, for distribution in the US alone. This is equivalent to 1 kina and 20 toea for every man and women in PNG over the age of 15.[1] A prominent statement of self-promotion to the larger world of global investment and finance, this advertising section is also a significant statement of national self-identification and aspiration.

---

1   The current estimated population of PPG is 8.3 million, of which 64.6 per cent are estimated to be more than 15 years of age.

Under the statement's section, 'Find a New Route to Prosperity', one reads that PNG is 'growing in stature as a global investment and tourism destination'. Under 'A Mine of Opportunity', it is proclaimed that 'the taxes and foreign currency that mining generates are … the engine of the country's development', and that the mining sector alone contributes more than 50 per cent of the country's entire export revenue (Eye on PNG 2018). In addition to existing mine sites such as Lihir and Porgera (Ok Tedi with its large-scale ecocide is not mentioned), new projects such as Frieda River, Wafi-Golpu and Ramu nickel and cobalt are foregrounded, along with Liquified Natural Gas (LNG) projects such as Hides, P'nyang and Juha. The advertising spread declares such projects to be the centrepiece of PNG's future development.

In recent years, as discussed further below, a range of anthropological research has explored the local, regional and national impact of resource extraction in PNG. This work has poignantly brought to light key issues concerning inequality, development and cultural response in and around major mining and oil/gas sites in PNG. Beyond these developments but importantly linked to them are the larger entrainments of expectation and disappointment among local peoples who are not directly impacted by large-scale resource extraction itself. These expectations and disappointments inform regional patterns of inequity and pursuit of development in its effective absence (see Knauft 2019a).

Across PNG, one finds enormous interest, attentiveness and preoccupation with the influence and potential impact of international resource extraction initiatives—sometimes all the more so because they have *not* materialised locally. In the Strickland-Bosavi area of PNG, these dynamics have been documented among the Kubo, where an LNG exploration camp was established for several years (see Minnegal and Dwyer 2017), and the Gebusi, who were subject to social mapping along the route of an anticipated LNG pipeline. In both cases, the promise of resource development is deferred despite great expectation and local excitement concerning major resource development projects elsewhere.

On a larger scale, the entire economy of PNG's Western Province has been severely impacted by—and continues to be dependent on—royalties from the Ok Tedi mine. As these revenues have reduced and dried up, along with the degradation of the Fly River ecosystem, government infrastructure outside the province's few towns has declined if not collapsed. This same trajectory characterises many rural outstations across PNG as a whole:

mines, LNG projects and other major resource extraction initiatives fuel the assumption that wealthy energy companies will build and maintain local infrastructure and services—while the government is absolved from responsibility even as it positions itself to receive the lion's share of royalties. In this respect, major resource extraction projects are in some ways all the more powerfully felt in and by their *absence*. More palpably, a sense of being left behind or left out is often at the heart of conflicts and disputes not only at or near the epicentre of resource extraction sites but in areas far distant.

I here examine resonating chains of expectation and inequity that devolve from major resource extraction projects in PNG. I consider these dynamics at resource extraction sites themselves and extend their trajectory to areas much less directly impacted, including as reflected in anticipatory hope, expectation and fanciful projection, if not fantasy. These processes both broaden and deepen our understanding of the larger dynamics and trajectories of perceived inequality that both connect and polarise peoples who are taken to benefit more, or less, from resource extraction in PNG. Given the anticipated centrality of large-scale resource projects in the economic and political future of PNG, these articulations seem particularly important.

In PNG, the impact of large-scale mineral and petroleum extraction projects presents not just a huge and consistently adverse impact on the communities most affected but a culture of expectation, frustration and disempowerment. There is a stark if not catastrophic or cataclysmic mismatch between plans and aspirations for resource extraction and the actual results of these projects in fomenting and escalating contention, inequality and misery. One cannot understand much less address this situation by considering economics alone, even in relation to politics. This is at heart a problem of *cultural* political economy, both national and local, including ideas, beliefs and values of modernity and progress against which local, regional and national realities become icons of failure and testaments to continued lack of development.

Conceptual and Theoretical Issues

An analytic issue from the start is our conceptual understanding of 'inequality', 'inequity' and 'culture'. Inequality tends to reference conditions of objective difference between individuals or groups of people, including especially their differences in possessing money or commodities and access to and exercise of political power. Inequity, by contrast, denotes the subjective sense of fairness or justice in these relations, especially the degree to which differences in wealth or political influence are seen as unjustified or intolerable. Though the two terms overlap and shade into one another in common usage, they beg different analytic and theoretical articulations. Documentation and analysis of inequality tend toward economic and political consideration of the material causes and conditions of differences in wealth and/or power. Considering inequity, on the other hand, leans toward understanding the subjective assumptions and dynamics whereby inequality is perceived and experienced as unfair and unjust. On the one hand, large differences of wealth and power may be accepted and legitimated by those with less as well as by those with more wealth. At the extreme, such differences may be seen as the natural social and cultural order of things. On the other hand, small or even negligible or imaginary differences of wealth or power may be seen as virulently unfair and unjust.

These differences are important, including in the specifics to be presently considered, insofar as they pinpoint the articulation or fulcrum point between differences of material condition and those of subjective perception or projection upon which social or political responses to inequality are formed and enacted. Melanesia in general and PNG in particular have long been considered areas in which strong ethics of egalitarianism make people vigilantly aware of and critically resentful of differences of material acquisition and wealth, especially insofar as such discrepancies are not based on differences in individuals' own work and physical labour (e.g. Read 1959; Sahlins 1963; see Knauft 1999: Ch. 1).

In a contemporary setting of mining and other forms of large-scale resource extraction, the slippage between an economistic assessment of rationally justified wealth differences and a subjective understanding of why people get so upset has large if not monumental consequences. It informs, for instance, the great differences of perception and power that inform mining officials, politicians and local people who benefit from major compensation vis-à-vis those many locals who receive no significant material payment.

Cultural processes are not just implicated in but are central to this disjunction. This is not to imply some unitary or bounded notion of culture. Rather it is to stress the subjective and perceptual dimension that so strongly informs people's experience. For present purposes, culture can be taken as the subjective dimension of social life, including especially its results in the collectivisation of subjectivity. As such, there is a cultural dimension to virtually everything in the social world, including the subjective experience of inequality and the attribution or projection of inequity. In the present context, this dynamic is at the heart of how and why Melanesian discontent is sown so deeply and spreads so far and wide in contexts of major resource extraction. These dynamics pertain not only to the sites of actual resource extraction but to much broader surrounding areas, especially including those that are not directly impacted by these projects in material terms. Such ostensibly non-impacted areas are often in fact profoundly influenced by the perception of large-scale inequality and the subjective experience of gnawing inequity.

A Cultural Conundrum

It is obvious that mining and oil/gas development in PNG have major negative consequences. These stretch from the disastrous Bougainville civil war following the bitter Panguna mining dispute with local people (May and Spriggs 1990; Liria 1993; Denoon 2000; Lasslett 2014), to the ecocide of major parts of the Fly River system from the Ok Tedi mine (Kirsch 2006, 2014, 2018), to the horrific violence and social degradation associated with the Porgera mine (Golub 2014; Jacka 2015), to rising tensions, inequality and restrictions of social networks at the 'best-case' offshore mine at Lihir (see Filer and Jackson 1989; Bainton 2009, 2010).

Beginning her paper on women and work in Lihir, Macintyre (2015: 1) writes:

> I asked a woman with whom I work: 'What does money do?' She replied 'It makes men drunk and young women single mothers—money has spoiled this place.' In 1994 Filer predicted various forms of 'social disintegration' for Lihir. Great economic inequalities that now exist, violent arguments, once rare, are commonplace. Millions of kina have been spent on beer. The simultaneous introduction of beer, roads, crimes and motor vehicles has its own devastating effect.

Concluding his book on the impacts of the Porgera mine, Jacka (2015: 231) states: 'In essence, I argue that Porgera is a massive development failure both socially and environmentally ... [T]he costs of mining in human lives and the degradation of biodiversity far outweigh the benefits of development.'

Jacka (2019) has recently documented the Porgera-inspired proliferation of a Rambo mentality, by which young men with expensive high-powered automatic rifles enter the so-called 'life market' to kill others repeatedly—until they themselves are killed in return. Reproducing the cycle, these deaths generate large-scale compensation demands and seed further conflict between the kin of the person killed and the killers. Absent effective compensation, the taking of further lives by way of revenge—negative reciprocity—is forcefully re-engaged.

Even among those privileged few at Porgera who receive major compensation benefits and relocation housing, Golub (2014: 139–40) suggests that their settlement was 'considered dangerous, dirty, degraded, and squalid', and so awash in drunkenness, gambling, prostitution, kept women and unsavoury and uncomfortable living conditions that many residents preferred to go and sleep in their traditional bush houses.

The problems associated with large-scale mining and oil/gas extraction projects are legion in PNG; they are practically a textbook case of the extractive resource curse in developing countries: lack of sustainable economic growth; dependency on unearned windfall profits; social and cultural degradation through alcohol abuse, sexual exploitation and gambling; and skyrocketing problems of national, provincial and local autocracy and corruption (cf. Ross 2001, 2013; Murshed 2018).

And yet, large new resource extraction projects remain the epitome of hope and positive promise in PNG among the large majority of politicians as well as the populace, including at national, regional and local village levels.

It is predictably hard to generalise about PNG, as it is about Melanesia, much less the wider Pacific Islands. It is a truism that PNG's renowned cultural diversity, including hundreds of local languages, makes it quite possibly the most culturally diverse country per unit land in the world (see Knauft 1999: Ch. 1). Anthropologically, Melanesia and New Guinea within it have long been considered the acid test case region for adjudicating general theories or ideas about pan-human psychology, social life and culture.

In this context, the project by Filer and Macintyre to survey the diversity of community responses to mining in Melanesia is particularly revealing. In case after case from across the country, they find that 'in the minds of most Papua New Guinean "grassroots" or village people, mining has become the way to gain wealth rapidly and to ensure that dreams of "development" and "modernity" come true' (Filer and Macintyre 2006: 216). Hence, even though non-governmental organisations 'have tended to emphasize the negative impacts of mining, especially on indigenous communities and their environments', they nonetheless find that the responses of Papua New Guineans themselves 'testify to the enthusiasm with which Melanesians welcome mining on their land' (ibid.: 221). Hence:

> [F]or a majority of Papua New Guinean villagers, the desire for development sweeps aside contemplation of its negative effects— even when these have been directly experienced. (ibid.: 223)

> While the reality might not conform to the myth-dreams of those who hope that their land will be the site for the next mine, there is sufficient evidence of relative wealth and advantage to feed aspiration among those who have no mine and nostalgia among those whose mine has closed. (ibid.: 224)

For all of mining's negative effects, then, we are left with a burning question: how and why are people up and down the social food chain, from grassroots villagers to national politicians, so fervently and petulantly desirous of these massive and challenging intrusions? And why, when the obvious results are so often so negative, do 'the experiences of marginality in one project in no way dampen enthusiasm for yet further large-scale projects' (Filer and Macintyre 2006: 226)? This issue is global and not limited to PNG. Chronicler of American society and culture, Arlie Hochschild (2018), finds a similar phenomenon among the large majority of Trump supporters in her native Louisiana. Despite enormous and crushing problems caused by petroleum industry refineries—including pollution and local ecocide, skyrocketing rates of environmentally caused cancer, and degrading and paltry employment prospects at the facilities themselves—residents support petrochemical industries on their doorstep and are loath to criticise them. The seeming explanation, in Louisiana as in many regions of PNG, is, simply, that there appears, at least, to be no other option.

The question of what constitutes a viable development option would appear on the surface to be an economic or at least politico-economic one, including the larger desire to secure more money and commodities. But the cultural assumption thereby skirted is that these will ultimately provide a more satisfying and happier life. As Weber (1958) and others have long suggested, capitalism is itself undergirded by cultural assumptions of value that are not given by economic rationality, but are themselves—when viewed in simple cost–benefit terms of work and reward—rather irrational. In the insular Pacific, mercurial and sometimes downright bizarre workings of Western capitalism are often dramatically on display—as Patterson and Macintyre effectively brought to light in their collection *Managing Modernity in the Western Pacific* (2011).

An important factor that informs these developments—from the capitalist sublime to the capitalist ridiculous—is the cultural idea that people deserve and are entitled to betterment by way of a new and better future (Koselleck 1985). This modern orientation, which has had strong impact in so-called marginal world areas such as Melanesia as well as in most other parts of the world, is thrown into relief by the region's diversity. A century or two ago, most New Guineans had no expectation of future betterment over time, no notion that time should unfold as an arrow of continuing progress into an unknown but hopeful future. Now, however, the cultural mandate for betterment and progress makes even the most dismal and difficult legacies of mining and its extractive cousins not just acceptable but practically necessary wherever it is financially and logistically feasible. There is a palpable sense that a lottery ticket for windfall betterment is worth almost any current risk or price. In the process, hundreds of New Guinea cultures have now become hundreds of inflections of localised and localising modernities of frustrated desire and entitlement, *not* giving *up* on so-called 'traditional' aspirations, but employing and inflecting local culture in and through the lens of needing and wanting progress, a better future and a more commodified way of life. Local versions of becoming or aspiring to become modern are if anything shot through, pervaded, with a depth and richness of local social and cultural resources bequeathed by long-standing practices, traditions and beliefs (see Knauft 2002a; cf. Knauft 2019a).

Porgera: The Golden Rainbow Goes Over the Hill — and Down the Other Side

An important pair of books throw the dynamics of cultural reaction and response to mining in PNG into special perspective: Golub's *Leviathans at the Gold Mine* (2014) and Jacka's *Alchemy in the Rainforest* (2015). Both pertain to the mega-mining development at Porgera in the Enga Province of the PNG highlands. But the books are as different, and as wonderfully complementary, as one could imagine. Golub, who is something of a Sahlinsian post-structuralist, lived and worked among core royalty-receiving landowners in a relocated clan settlement near the epicentre of the mine—located on 'a bulging pocket of land bounded by the waste dump [of the mine] on two sides and the open pit of the mine on a third side' (2014: 136). His study focuses on the dynamics whereby Ipili core landowners, on the one hand, and 'the mine', on the other, became reified and powerful Leviathan-like entities of legal, political and economic power, prominence and contestation. He also notes in an afterword how brutally conditions had further declined after he left in 2001 and returned six years later—drunkenness, crime, enormous illegal in-migration, violence and spiralling claims of victimisation, shooting and bad faith on opposed sides. Indeed, conditions became so problematic that Golub himself could no longer straddle the reified division between the Ipili and the mine that he so effectively describes (ibid.: 210). Golub started his work hoping Ipili would be a success story showing how assertive indigenous people and a major mining company could negotiate arrangements to their mutual benefit (ibid.: 212). He ends up believing that the valley would have been better off if the mine had never been built, concluding that: 'The Porgera experiment is over, and the Ipili are the losers' (ibid.: 213).

Jacka's book, as he notes (2015: 10), contrasts with Golub's in being a detailed study of Ipili who are *not* primary landowners of the mine site—but who were promised various forms of development programs. They have been wildly excited about these as part of the penumbral mine benefits promised by the company and the government. Bringing to bear interest and expertise in ecology, forestry and subsistence livelihood in addition to social and cultural change, Jacka documents how for most Ipili, the elite landowners and their wealth are a source of great desire and envy and also of resentment by others. Gravitating to the high-altitude road that leads directly to Porgera, many Ipili, and especially young men, have compromised their subsistence—along with their more fragile, higher-elevation environment—by adopting a 'highway life' of 'doing nothing' and 'wandering' (*raun raun*) in aimless search of wealth and excitement (ibid.: 203ff.). Resentment festers between them and the lower-dwelling Ipili, who consider life along the highway degraded, immoral (including rampant prostitution) and without proper social relations of exchange and reciprocity. In the mix, non-compensated Ipili have become bitterly embroiled in armed disputes among themselves over the lack of resources and failure of development projects; their disputes have become locked in a vicious circle as both cause and effect of failed development.

Project infrastructure has been completely destroyed by local fighting, along with wholesale burning and looting of houses, stores and practically all other structures. Killings, revenge killings and warfare with high-powered rifles escalated to the point that homicide compensation for the many persons slain rose to astronomical levels—forestalling restitution and preventing reconciliation (see Jacka 2019).

These problems are thrown into stark relief insofar as they persist in the shadow of others' excessive and profligate wealth. While most Ipili remain subsistence farmers, new 'super big men' may drive very expensive cars and have as many as 25 or 30 wives or kept women sequestered in dormitory-like compounds (Jacka 2015: 210). As is extremely common in major cases of mining compensation payments, core landowners tightly restrict and narrow their traditional social networks in order to retain their wealth. This pattern is particularly striking among Ipili, whose pre-mining system of kinship, exchange and reciprocity was especially extensive and flexible (Golub 2014: 113–33; Jacka 2015: Pt II). Against this, enormous immigration by those from other areas, attracted by Porgera's wealth, has allowed a range of them to become insinuated by marriage, co-residence or gifts into the compensation payout system—a system from which most Ipili, who are not landowners at the mine site itself, are otherwise excluded. As such, mining-induced disputes, cleavages and factionalisation within Ipili society as described by Jacka articulate with and echo the escalating elite polarisation between legally reified Leviathans as described by Golub—between the mining corporation and the select and enormously wealthy few Ipili landowners of the mine land itself.

This cycle of contestation, factionalisation and polarisation is not limited to one subgroup or dimension of the sprawling social and economic inequalities fomented by the Porgera mine; rather, it ramifies extensively. For instance, the Huli people, some 200 kilometres from the mine site and in a wholly different province, have significant relations of clientage and wife-giving to wealthy Porgeran men. Wardlow (2019) describes how a range of Huli women have been sold off for elevated bride price to Porgera men, only to find that their lives were not luxurious and free of work but degraded as servants and sex objects—reduced in effect to sexual and domestic slavery. Some of them even considered it a positive relief to have contracted HIV from their husbands and hence be able to leave Porgera and return to their Huli homeland—as titled in Wardlow's (2019) article, 'With AIDS I am Happier than I Have Ever Been Before'.

In essence, Golub's *Leviathans at the Gold Mine* focuses on the extravagant, divisive and ultimately debilitating and degrading impacts of mining compensation among those few who qualify as local landowners at the mine site itself. Jacka's *Alchemy in the Rainforest,* by contrast, focuses on the effect of resentment, jealousy and aspiration that leads to internalised conflict, devastating warfare and destruction among those who are *not* direct mine site landowners but who live in the general area. Finally are accounts by Wardlow describing the *indirect* impact of the Porgera mine among the Huli, across a provincial boundary.

The larger point is that a cultural economy of inequity spirals desire, envy, aspiration and resentment. The experience of inequity does not stop at the border of the mine itself, nor even at the border of the larger ethnic group within which it is situated. Rather, the wealth associated with the mine becomes alternately a magnet of attraction from elsewhere and a lightning rod for dispute. Attraction from afar is driven by modern aspiration—a desire to be better off along a newly demonstrable yardstick of unimaginable wealth. Against this, internal disputes are driven by the spectre or the reality of wealth and possessiveness that lie newly outside the bounds of long-standing norms, values and expectations of exchange, sociality and reciprocity (cf. Strathern 1988; see Gregory 2015). The abrogation of meaningful reciprocity is often at the cultural heart of Melanesian tensions that intersect with and inflame desires and resentments of being or becoming modern. This is a widespread pattern. Among the Yonggom people subjected to ecocide of their environment from the tailings of the Ok Tedi mine, development is perceived simply and revealingly described as 'failed exchange' (Kirsch 2006: 95).

The Strickland-Bosavi area: 'Of Course it Might', OR, Things in the Mirror May Be Closer Than They Appear

What about groups yet further afar, beyond even the penumbra of economic and demographic connections with the mine site itself? Here issues of anticipation, expectation and projection come strongly into play, including in very remote areas that are hardly on the map of anyone's scheme of development or modernity—except in the perception of local people themselves. This perspective foregrounds the impact of mining and oil/gas developments not by their presence but their crushing absence.

In their innovative article, 'Waiting for Company', Dwyer and Minnegal (1998) describe the poignancy of the remote and distant Kubo people, who reside in PNG's Western Province in the northwesternmost section of the Strickland-Bosavi area. As described by Dwyer and Minnegal (1998) Kubo wait patiently, in principle endlessly, for outside agents, institutions and corporations to come and lift them up, empower them, give them development and fulfill their aspirations.

> All outsiders have failed Kubo. At each small community the people's explicit complaints are the same. The missionaries have not come, nor have they sent pastors. The White missionaries based at larger communities are themselves departing. The government has not provided school or aid post or funds for the construction of an airstrip; nor does it offer more than an occasional 'make-work' programme. Mining companies are transient, employing people for a few months and then departing with no guarantee of return. The rumours concerning logging companies are just that; rumours that encourage an understanding that it is only people elsewhere who receive the benefits of modernity. And the anthropologists are too few in number. (Dwyer and Minnegal 1998: 32)

Decades later, Kubo are still waiting (Minnegal and Dwyer 2017). A local Esso seismic survey and alluvial gold search in 1985/86 raised Kubo expectations astronomically. But by 1987 Esso had left. For months, even years, their left-behind equipment and stores were painstakingly relocated and stored by Kubo. But Esso never returned. By 1995, Kubo were still 'waiting for company', complaining that outsiders had not come and accumulating a lengthening list of disappointments. They were ecstatic in early 1996 when Porgera reconnaissance helicopters landed, giving each village 200 kina to build a helipad. But then they departed, and after a few more furtive visits, never returned. By the 2000s, natural gas development was actively pursued among the Febi people just north of the Kubo. Five gas wellheads were drilled at the associated Juha gas site. Associated with this, a significant exploration base camp was established next to the Kubo's airstrip, at Suabi. Kubo began to associate more intensely with Febi in hopes of identifying with them for compensation. In 2014, the multi-billion-dollar PNG LNG project shipped its first natural gas from another pipeline, through 700 kilometres of rainforest from interior New Guinea all the way to the coast—but not from areas near to the Kubo.

> People at Suabi had high hopes for future benefits from the PNG LNG Project in the form of royalty payments and business development grants. In their understanding, those benefits would

> be provided, either directly or indirectly, by 'Company'. And, for
> them, Company had a very material, and personalised, presence
> (compare Golub 2014); a presence that offered opportunities in
> the present for those who were able to discern and act on them.
> The camp at the Suabi airstrip was woven into everyday life.
> (Minnegal and Dwyer 2017: 14)

The LNG base camp at Suabi was active for a range of seismic exploratory
projects from late 2012 to early 2014. Kubo drew up elaborate lists
and complex calculations concerning who among them would get
compensated and by how much (Minnegal and Dwyer 2017: 144ff.).
But at the same time that the Hides gas pipeline started piping its mega-
exports, Esso decided to close and dismantle its small exploratory base
camp in Kubo territory. Further, the Juha wellheads in Febi territory just
north of the Kubo remained offline and unconnected to the extensive gas
pipeline further east; they remain on hold and not scheduled to produce
until the mid-2020s—or perhaps indefinitely if the global price of liquid
natural gas does not increase. To local people, this further and potentially
indefinite delay is extremely frustrating; their hopes are dashed yet again
(ibid.: 78). Adding to the confusion was a social mapping survey that
documented residence rights for land compensation associated with yet
another projected pipeline that would connect with a major gas wellhead
yet further northwest, at P'nyang. But the projected pipeline corridor
land compensation area edged just outside of all Kubo territory—and the
pipeline was never built in any event.

> From the perspectives of Kubo and Febi people, through these three
> decades … explorers … found valuable resources and held them
> for the future. There was, it seemed, much secrecy. Eventually,
> however, those resources would be taken from the ground and
> royalties paid to land owners. In the meantime it was necessary
> to host the company representatives and, as possible, accept
> employment at base camp or in the field as labourers, assistant
> loadmasters, security officers, laundry workers, assistant cooks
> and so forth. Only one Kubo man had permanent employment
> associated with exploration activities, as a fully trained loadmaster
> with Pacific Helicopters. Kubo people, in particular, felt
> disenfranchised. (Minnegal and Dwyer 2017: 80)

In December 2013, Kubo finally did receive some money—a government
infrastructure grant of K81,500 to support maintenance work on the
airstrip and to fund the building of a new community health centre
(Minnegal and Dwyer 2017: 167). But a significant portion of the money

'disappeared', and a divisive court case ensued. In 2014, representatives of Talisman—the company that had taken over the exploration camp at Suabi, provided what they considered to be a 'final payment' to local people—in the very modest amount of K6,400. The distribution of this money among Kubo was fraught and contentious. The exploration camp was dismantled and shut down, with remaining material goods given out until nothing was left.

> To people at Suabi ... the present is deeply imbued with desires that are oriented towards a future, a future in which 'development' comes, in which they are no longer 'remote' and forgotten, and in which wealth that is perceived as 'rightfully' theirs is given material expression either in the form of extractable resources—gas, oil, minerals, timber—on their own lands or rights to the benefits expected from such resources on the lands of their immediate neighbours. (Minnegal and Dwyer 2017: 205)

Over decades, Kubo are still waiting for their company. And through waiting, their social relations, their sense of money and individuality in lieu of community integrity have fundamentally changed.

Further Down the Line ...

Gebusi and their direct neighbours are yet lower in elevation and of less interest geologically than even the Kubo or the Febi. But this has not lowered the significance of the presence of the absence of development; rather, it has underscored it. Focusing on issues of development and resource extraction, there might seem little to say about Gebusi (cf. Knauft 1985, 2002b). And yet Gebusi are highly interested in and desirous of the kind of development they associate with Ok Tedi to the northwest, Juha to the north, Hides gas to the east and Porgera to the northeast. None of these project sides can be viably accessed by Gebusi, as this would require weeks of trekking, dangerous crossings of major rivers and traversing the country of multiple foreign or enemy groups. There are no roads within their own Nomad Sub-District (see Figure 6.1). But one edge of their territory is within the 5-kilometre border of the projected route of the P'nyang gas pipeline after it is tunnelled beneath the Strickland River— though it remains highly doubtful if this will ever happen in fact.

Figure 6.1 Untrammelled Gebusi rainforest, with Mt Sisa (left) and Mt Bosavi (far right) in the distance.
Source: Photo by Bruce Knauft.

In 2013, I charted the borders of Gebusi tribal territory by GPS and wrote and publicly posted a report about customary Gebusi land tenure practices (Knauft 2013). This was done to help ensure that if the pipeline *was* ever built there would be independent documentation of—and hopefully at least some protection of—Gebusi land rights. But describing a gas pipeline in vernacular Gebusi is almost impossible. When I gave one high school leaver who didn't know me very well my business card, he saw I was from the Emory College of Arts and *Sciences*. Surmising I was a scientist, he proceeded to ask if I was taking up Gebusi minerals into my GPS so I could take them away and sell them.

When a social mapping helicopter did in fact circle the village to pay me a field visit, residents went wild. In a frenzy, several young men even grabbed axes and tried to hack down venerable coconut trees that ringed the village clearing so the chopper could land more easily. As the chopper fluttered away to land in a larger clearing by the mission station, everyone chased after it. Though my brief meeting with the social mapping representative was inconsequential (see Knauft 2016: 201–8), it was momentous for Gebusi. It put them on the map: of all the places the helicopter flew over, the one spot it chose to land in their whole area was on their doorstep. This validated for them, in a way that I never could, the legitimacy of—and expectation of benefit from—the GPS mapping that I was doing.

As described elsewhere (Knauft 2019b), Gebusi have a virtually wage-less economy. Yet, in the absence of that, they have developed elaborate standards and records of daily work for a host of activities that in principle could, but almost never are in fact, paid for in money.

Likewise, Gebusi are highly attuned to the potential value of their land—if and when outsiders ever take an interest in it. When reports circulated of social disruption around the Hides gas project station at Moro, rumours circulated that the Nomad Station was being considered as the location of an alternative LNG field office site. The Nomad airstrip would be repaired and enlarged, cell phone towers would be built and made working, new government offices would be built, the school would be upgraded, and the Nomad market would flourish once again. This prospect generated wild and inflated enthusiasm in large part for the very reason that the Nomad Sub-District government offices were all closed, the airstrip is now closed, as well and the existing cell phone tower is non-functional, with no plans for its repair (see Figure 6.2).

Figure 6.2 Abandoned government house at Nomad Station, 2013.
Source: Photo by Bruce Knauft.

Conflict and Violence Fuelled by Lack of Development

As among Kubo, Gebusi heightened their commitment to be ready for modern development whenever it might eventually come. They drew up elaborate lists of landowners and charted land boundaries, especially those at or close to the Nomad Station, where land was expected to be dear. Long-simmering disputes escalated between clan members who claimed ancestral rights to part of the land near the Nomad airstrip.

Across the Strickland River, at a place called Yebo (Yavo), stories surfaced in 2016 that the Talisman exploration company would be building a major wharf and staging port for its LNG exploration teams to the north. The few people in Gebusi territory who had ties to this area rushed across the river and tried to legitimise land claims to the locale in question. In the process, they spent large amounts of very scarce money in an attempt to get their claims legally processed in Kiunga for contestation in court. These attempts were predictably futile; they lost their money in legal fees, and the case was never brought to trial. Their attempt nonetheless caused enmity not only between the Gebusi residents and their Strickland kinsmen but between them and others in their home settlement at Gasumi Corners. Those long resident in the community claimed that the attempt by their co-residents to establish land rights on the other side of the Strickland River reflected and reinforced their non-local identity—and undercut their right to continue residing in Gasumi Corners itself. When the senior man of the group pursuing land rights in Yebo died, fears mounted that tensions between his clan and the rest of the settlement would erupt in sorcery accusations and split the community.

Even in the absence of any development at all, then, its possibility, its potential presence, is enough to fuel expectations, plans and intensifying disputes. As in the present case, all this can occur without a single kina being given or even asked for in compensation.

A yet more dramatic case is that of Powa, a senior Bedamini man in a remote mixed Bedamini-Gebusi village who was tied to a tree and executed as a sorcerer in May 2016 (see Knauft with Malbrancke 2022: Ch. 4). The rationale for Powa's execution, carried out collectively by Bedamini through a large network of persons, was that he had worked magic against his wife's son from her previous marriage in revenge for a presumed land dispute between them. It was reasoned that the son was a natal owner of local land that might, at least in hypothetical principle, be subject to eventual compensation for LNG pipeline passage by ExxonMobil. Wanting to take over this young man's presumed land claim, it was retroactively believed that Powa, as the man's stepfather, had killed him to arrogate the land claim onto himself. As I discerned during an ethnographic visit to the distant village, there was no evidence that any compensation funds had been in any way promised or even hinted at by ExxonMobil, no evidence that the pipeline would ever be built, and not even evidence or history of an open or persisting dispute over the land issue beforehand between Powa and his stepson: it was largely if not completely projected post-factor. But in the context of a young man's sudden and unexpected death, and projected need and greed in the complete absence of resource extraction development and compensation, Powa was presumed to have killed his stepson by sorcery and was executed by his Bedamini relatives in return.

In their own way, Gebusi, just next door, are already in the cultural economy of large-scale resource extraction; it has already changed their calculus of action and expectation. They are influenced in concrete material terms as well. Gebusi have no cash crops or other resources that are valued by outsiders. The sole exception is marijuana, which is grown and traded surreptitiously. At great personal risk, this contraband can be carried across the territory of several different ethnic groups and sold in or near the PNG highlands. The other commodity, even more risky, is the rumoured existence of an overland human transport network that ferries disassembled high-powered rifles up the Strickland River from Australia across the Torres Strait and via the Strickland-Bosavi area to highland areas such as Porgera. In a world of no money, running drugs and guns to mining and gas areas that are flush with cash has become the only viable option for significant Gebusi 'development', as risky and deplorable as Gebusi find these activities to otherwise be.

In recent years, the Nomad airstrip has closed and virtually all government services have ceased; as Gebusi put it, 'The government has died' (*gamani golom-da*). Gebusi still have no roads to anywhere and virtually no wage economy. The average adult daily income of between USD0.10 and USD0.20 a day is between one-tenth and one-twentieth of the absolute world poverty level of USD2.00 per day (see Knauft 2019b).[2] Like Kubo, Gebusi have been tantalised for decades by the possibility but the continuing absence of any resource development or compensation (see Figure 6.3).

---

2   For a field video of underdevelopment among Gebusi, see: www.youtube.com/watch?v=SqSf7X yJbHs

And yet, as a virtual replacement for this non-economy, Gebusi have developed robust and elaborate time sheets and record-keeping for a host of activities that could in principle be paid (Knauft 2019b).

Figure 6.3 A Gebusi man wearing ragged clothes, 2017.
Source: Photo by Bruce Knauft.

Reactive Modernity

It is perhaps unsurprising that studies of large-scale resource extraction projects in world areas such as Melanesia focus on the land and the people most directly affected. This includes indigenous discontent, resistance or violence near and along the path of resource extraction. But this seems just the tip of the iceberg. Under conditions of locally modern aspiration, mining or LNG developments can be as or even more important by their local absence as by their presence.

Major resource sites of mega-money, machines and men are certainly a magnet of attention in countries like PNG; their travelling imaginary of good fortune and amazing wealth spreads far and wide (cf. Mageo and Knauft 2020). Among rural peoples such as the Gebusi and Kubo, however, actual information and accurate news pales beside the force of cultural and imaginative projection. This process is hardly new. In 1981, I trekked to a distant Gebusi village to attend the spirit séance of Wahiaw, a renowned shaman. Crippled for many years, Wahiaw could not walk and had to be carried from settlement to settlement; it is highly unlikely he ever saw the Nomad Station or its officers, as his remote settlement was days' walk from the post. Yet his spirit séances were flush with fantastic stories, allegories and dramas—often quite accurate in spirit—told by the spirit people (*to di os*) about a spirit-world incarnation of outside officials, work projects and fantastic wealth (see Knauft 1989). Bordering on cargoist in zeitgeist, Wahiaw's narratives were at the same time cautionary tales about the excesses, impersonality and lack of reciprocity in money-fuelled development.

Over the years, however, the kind of caution voiced by Wahiaw has largely fallen by the wayside. Not that Gebusi have much reliable information about developments outside their own narrow range of direct experience. But their aspirations can be fuelled all the more by lack of reliable information or reality checks. Even by 1998, Nomad schoolboys regularly envisaged themselves as successful workers at the distant Ok Tedi mine. As reflected in a drawing by Tony Semo, they envisaged a world of plenty in which the forest was also brimming with wildlife and a smiling sun (see Figure 6.4).

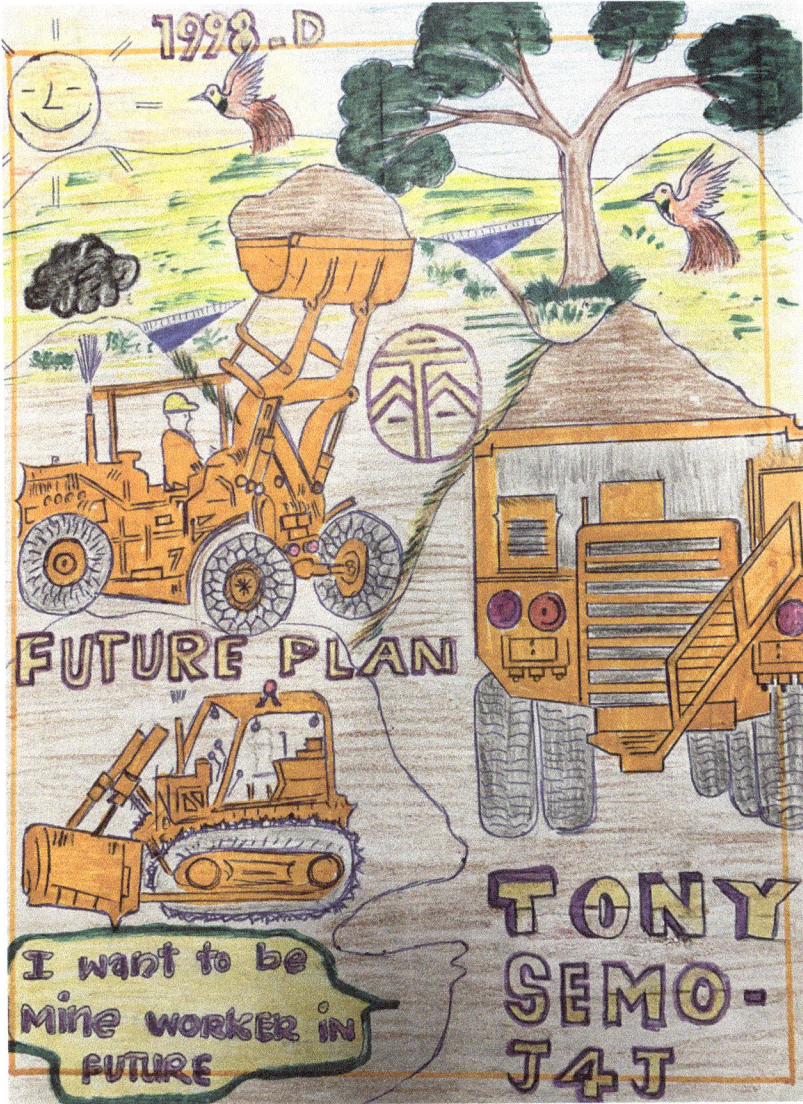

Figure 6.4 Schoolboy drawing in 1998 by Tony Semo—of aspiring to be a successful heavy equipment operator at the Ok Tedi mine.

Source: Photo by Bruce Knauft.

Here we confront modernity, the time–space projection of beneficial progress into an unknown but desired future (cf. Harvey 1989; see Knauft 2002a: Introduction). This is a future that is almost invariably projected as the alter or antithesis of traditional culture, which by contrast is easily seen as deficient or backward (for Gebusi, see Knauft 2002a: Ch. 3). In PNG, progress is iconically if not overwhelmingly associated with major mining and oil or gas projects; these are taken as the armature for social and economic 'development' across the country as a whole.

In this cultural context, the impact of major resource extraction projects does not decrease as a function of geographic distance. Rather, cultural exposure and relative deprivation create alternative or opposed logics whereby the impact of 'development envy' can be as great or even greater among those further away. Their motivation and projection can be all the stronger given their remoteness and difficulty of access. The illusion that the impact of major resource extraction projects is at the centre, radiating out, betrays a bias shared by energy companies and the PNG state—namely, that problems outside a very limited area of direct impact lie outside their responsibility or concern. Indeed, it was highly evident during work among Gebusi in 2016 and 2017 that the government has given up any attempt to develop infrastructure in the Nomad Sub-District, as the area is considered to have no exploitable resources.

Given this, and notwithstanding the calamity of Porgera, new projects such as the potential P'ynang LNG pipeline bring heightened attention and hope to Gebusi, Kubo and other peoples of PNG's Western Province. Yet the triangular structure of relationship between landowners, multinational energy corporations and the government of PNG practically ensures these aspirations will result in divisive conflict and antagonism even in a best-case scenario.

Key here is the issue of land. In an attempt to shield energy companies from local discontent, it is now increasingly mandated by the PNG Government that issues of land ownership in areas of potential compensation be settled in court (see Church, this volume). In effect, landowners must negotiate via the highly problematic legal mechanism of the PNG state rather than having their claims made or adjudicated by the resource extraction company itself. Government officials increasingly insist that compensation packages cannot be implemented until it is definitively determined who has a clear and confirmed title to each portion of relevant land. This, in turn, cannot occur until all disputes concerning land ownership and boundaries have been settled by the disputants themselves in court. As such, the onus of legitimate representation devolves from the grand collective indigenous Leviathan as depicted by Golub (2014) to the individual landowner, who must pay (and bribe) to receive a confirmed land title. This affords increasing leverage to those who already have money and to those who ally with wealthy, if often unscrupulous, outsiders. The circle is thus easily closed between wealth obtained by government corruption or graft and the ability to further increase wealth by buying into or paying off those in land compensation cases. In the process, local landowners lose their direct bargaining power and are forced to negotiate vis-à-vis a largely corrupt legal system and a highly aggressive rent-seeking state. The way that those with money can effectively buy into and exploit this process is prominently evident in the accounts of Jacka (2015) and especially Golub (2014) concerning the Porgera mine.

Conclusion

Large-scale resource extraction projects in PNG have an enormous cultural as well as economic impact in the minds of people in remote rural areas as well as across the nation. Against the unearned largesse of major land compensation, almost any local development scheme can seem, by contrast, to be a two-bit ante, hardly worth the effort. Instead, the symbol and significance of the mega-development site becomes a great looming imaginary, potent and powerful by virtue of its very distance from rural realities. This exacerbates a deep sense of inequity and resentment both at the local level and across larger regional and national networks and constituencies. The enormous gap between expectation and actual result makes these tensions ripe for generating conflict.

In such contexts, social fragmentation and divisiveness through aspiration and competition can continue in a reinforcing cycle. The Ipili mine landowners at Porgera contest against the Porgera Joint Venture Company. The Ipili *non*-mine landowners contest against the Ipili elite. Within the Ipili non-elite, those living in villages contest against those living along the Porgera road. On a broader scale, the Huli curry favour with but resent the Ipili. The Kubo resent the Febi. And, if either of them were to get compensation, the Gebusi would resent them as well. Even among Gebusi, with no compensation or viable prospect thereof, resentments arise between those who *might* be able to claim compensation *if* development comes, and those who presumably or ostensibly cannot. As described further above, an elderly Bedamini man was crucified in 2016 based on the belief that he sent sickness that killed his stepson (see Figure 6.5).

Figure 6.5 Enemies of the slain Bedamini sorcery suspect pose with their weapons as proud killers.

Source: Photo: Bruce Knauft.

Most people in rural PNG fantasise and project, but more practically and immediately they yearn. They yearn for the amazing wealth they perceive in other areas beyond their own. This underscores rather than diminishes the monetary value they perceive in the one potentially valuable commodity that they have: their land. As such, conflicts easily arise over the hoped-for benefit that development could hypothetically bring even in areas such as Gebusi where there is no evidence of land shortage or overpopulation. Hence we find reinforcing cycles between the projection of fantastic windfall, disappointment, waiting and a reinforced sense of being left out. This feeling of being unfairly deprivileged is magnified against the imagined shining gem of almost unimaginable wealth projected over the hill or even in the hoped-for future of one's own village. Absence of resource development does not forestall or efface such effects but can easily magnify them. This is not the abjection of *develop-man* in Sahlins's (1992) sense of the term (cf. Robbins and Wardlow 2005). Rather, it is a local inflection or incarnation of being a suffering subject, of being left behind in the tidal development of modernity—and precluded from benefit by others who stand in one's rightful way (see Robbins 2013; cf. Knauft 2019c).

Against the elevated and impossible standard of gargantuan riches, everyone can feel disadvantaged and left out—relative to those known or believed to have received so much more. Development projects such as Porgera, Lihir and Ok Tedi are indeed a success in at least raw economic terms for those very few who receive a disproportionate share of wealth. For others, including those scores or even hundreds of miles away, mega-development projects are a shining symbol of one's own backwardness. In this sense, large-scale resource extraction projects in PNG and other parts of Melanesia are the pinnacle not only of fantasised projection but also of ultimate inequity bequeathed by modern development.

References

Bainton, N.A., 2009. 'Keeping the Network Out of View: Mining, Distinctions and Exclusion in Melanesia.' *Oceania* 79(1): 18–33. doi.org/10.1002/j.1834-4461.2009.tb00048.x

———, 2010. *The Lihir Destiny: Cultural Responses to Mining in Melanesia.* Canberra: ANU E Press (Asia-Pacific Environment Monographs). doi.org/10.22459/LD.10.2010

Denoon, D., 2000. *Getting Under the Skin: The Bougainville Copper Agreement and the Creation of the Panguna Mine.* Melbourne: Melbourne University Press.

Dwyer, P.D. and M. Minnegal, 1998. 'Waiting for Company: Ethos and Environment Among Kubo of Papua New Guinea.' *Journal of the Royal Anthropological Institute* 4(1): 23–42. doi.org/10.2307/3034426

Eye on PNG, 2018. 'Eye on Papua New Guinea: Find a New Route to Prosperity.' *The Wall Street Journal.* Special Advertising Supplement. 6 November.

Filer, C. and R.T. Jackson, 1989. *The Social and Economic Impact of a Gold Mine on Lihir* (2 volumes). Port Moresby: The Lihir Liaison Committee.

Filer, C. and M. Macintyre, 2006. 'Grass Roots and Deep Holes: Community Responses to Mining in Melanesia.' *The Contemporary Pacific* 18(2): 215–231. doi.org/10.1353/cp.2006.0012

Golub, A., 2014. *Leviathans at the Gold Mine: Creating Indigenous and Corporate Actors in Papua New Guinea.* Durham: Duke University Press. doi.org/10.1515/9780822377399

Gregory, C.A., 2015. *Gifts and Commodities* (2nd edition). Chicago: Hau Books.

Harvey, D., 1989. *The Condition of Postmodernity: An Enquiry into the Origins of Cultural Change.* Cambridge: Blackwell.

Hochschild, A.R., 2018. *Strangers in Their Own Land: Anger and Mourning on the American Right.* New York: The New Press.

Jacka, J.K., 2015. *Alchemy in the Rain Forest: Politics, Ecology, and Resilience in a New Guinea Mining Area.* Durham: Duke University Press. doi.org/10.1215/9780822375012

———, 2019. 'Resource Conflicts and the Anthropology of the Dark and the Good in Highlands Papua New Guinea.' *The Australian Journal of Anthropology* 30(1): 35–52. doi.org/10.1111/taja.12302

Kirsch, S., 2006. *Reverse Anthropology: Indigenous Analysis of Social and Environmental Relations in New Guinea.* Stanford: Stanford University Press.

———, 2014. *Mining Capitalism: The Relationship Between Corporations and Their Critics.* Berkeley: University of California Press. doi.org/10.1525/9780520957596

———, 2018. *Engaged Anthropology: Politics Beyond the Text.* Berkeley: University of California Press. doi.org/10.1525/california/9780520297944.001.0001

Knauft, B.M., 1985. *Good Company and Violence: Sorcery and Social Action in a Lowland New Guinea Society.* Berkeley: University of California Press.

———, 1989. 'Imagery, Pronouncement, and the Aesthetics of Reception in Gebusi Spirit Mediumship.' In M. Stephen and G.H. Herdt (eds), *The Religious Imagination in New Guinea.* New Brunswick: Rutgers University Press.

———, 1999. *From Primitive to Postcolonial in Melanesia and Anthropology.* Ann Arbor: University of Michigan Press. doi.org/10.3998/mpub.10934

———, 2002a. *Critically Modern: Alternatives, Alterities, Anthropologies.* Bloomington: Indiana University Press.

———, 2002b. *Exchanging the Past: A Rainforest World of Before and After.* Chicago: University of Chicago Press.

———, 2013. 'Land Ownership in the Near-Western Nomad Sub-District, Western Province, Papua New Guinea.' Report. Viewed 15 May 2020 at: scholarblogs.emory.edu/bknauft/files/2015/07/Gebusi-Land-Report.pdf

———, 2016. *The Gebusi: Lives Transformed in a Rainforest World* (4th edition). Long Grove: Waveland.

———— (ed.), 2019a. 'Good Life in Dark Times? Melanesian Interventions in Dark Anthropology/Anthropology of the Good.' *The Australian Journal of Anthropology* 30(1) (Special Issue).

————, 2019b. 'Finding the Good: Reactive Modernity among the Gebusi, in the Pacific, and Elsewhere.' *The Australian Journal of Anthropology* 30(1): 84–103. doi.org/10.1111/taja.12303

————, 2019c. 'Good Anthropology in Dark Times: Critical Appraisal and Ethnographic Application.' *The Australian Journal of Anthropology* 30(1): 3–17. doi.org/10.1111/taja.12300

Knauft, B.M. with A.-S. Malbrancke, 2022. *The Gebusi: Lives Transformed in a Rainforest World* (5th edition). Long Grove: Waveland.

Koselleck, R., 1985. *Futures Past: On the Semantics of Historical Time.* Cambridge: MIT Press.

Lasslett, K., 2014. *State Crime on the Margins of Empire: Rio Tinto, the War on Bougainville and Resistance to Mining.* London: Pluto Press. doi.org/10.2307/j.ctt183p781

Liria, Y.A., 1993. *Bougainville Campaign Diary.* Victoria, Australia: Indra Publishing.

Macintyre, M., 2015. 'The Changing Value of Women's Work on Lihir.' The World Bank. Viewed 24 January 2022 at: www.researchgate.net/publication/265822576_The_Changing_Value_of_Women%27s_Work_on_Lihir

Mageo, J. and B. Knauft (eds), 2020. *Pacific Island Encounters: New Lives of Old Imaginaries.* Oxford: Berghahn.

May, R.J. and M. Spriggs (eds), 1990. *The Bougainville Crisis.* Bathurst: Crawford House.

Minnegal, M. and P.D. Dwyer, 2017. *Navigating the Future: An Ethnography of Change in Papua New Guinea.* Canberra: ANU Press (Asia-Pacific Environment Monographs). doi.org/10.22459/NTF.06.2017

Murshed, S.M., 2018. *The Resource Curse.* Newcastle, UK: Agenda Publishing. doi.org/10.2307/j.ctv5cg8kq

Patterson, M. and M. Macintyre (eds), 2011. *Managing Modernity in the Western Pacific.* St. Lucia: University of Queensland Press.

Read, K.E., 1959. 'Leadership and Consensus in a New Guinea Society.' *Man* 61: 425–436. doi.org/10.1525/aa.1959.61.3.02a00060

Robbins, J., 2013. 'Beyond the Suffering Subject: Toward an Anthropology of the Good.' *Journal of the Royal Anthropological Institute* 19: 447–462. doi.org/10.1111/1467-9655.12044

Robbins, J. and H. Wardlow (eds), 2005. *The Making of Global and Local Modernities in Melanesia.* Burlington: Ashgate.

Ross, M.L., 2001. *Timber Booms and Institutional Breakdown in Southeast Asia.* Cambridge: Cambridge University Press. doi.org/10.1017/CBO9780511510359

———, 2013. *The Oil Curse: How Petroleum Wealth Shapes the Development of Nations.* Princeton: Princeton University Press. doi.org/10.1515/9781400841929

Sahlins, M.D., 1963. 'Poor Man, Rich Man, Big-Man, Chief: Political Types in Melanesia and Polynesia.' *Comparative Studies in Society and History* 5: 285–303. doi.org/10.1017/S0010417500001729

———, 1992. 'The Politics of Develop-Man in the Pacific.' *RES* 21: 13–25. doi.org/10.1086/RESv21n1ms20166839

Strathern, M., 1988. *The Gender of the Gift: Problems with Women and Problems with Society in Melanesia.* Berkeley: University of California Press. doi.org/10.1525/california/9780520064232.001.0001

Wardlow, H., 2019. '"With AIDS I am Happier than I Have Ever Been Before."' *The Australian Journal of Anthropology* 30(1): 53–67. doi.org/10.1111/taja.12304

Weber, M., 1958. *The Protestant Ethic and the Spirit of Capitalism.* New York: Scribner's.

7

Reflecting on Resource-Driven Inequalities

Glenn Banks

Inequality does not follow a deterministic process. In a sense, both Marx and Kuznets were wrong. There are powerful forces pushing alternately in the direction of rising or shrinking inequality. Which one dominates depends on the institutions and policies that societies choose to adopt. (Picketty and Saez 2014: 842–3)

Introduction

Inequality is a central concern of economics and the social sciences across most of the West right now. The statement above by economists Thomas Picketty and Emmanuel Saez points to questions beyond an assumed naturalism or inevitability in the development of inequality. They make a case for the significance of deliberate policy interventions and of existing institutions in shaping the development of novel forms of inequality within societies. Examination of instruments such as tax regimes and interventions such as social protection networks in many parts of the world clearly provides support for this argument. In Melanesia, though, where state policy and presence is much less overt in people's lives, there is less evidence that formal policy and institutions are the dominant influences on the rise or fall of inequality. Indeed, what limited work that has been done points to a remarkably stubborn level of inequality (Bainton and McDougall 2021), despite changes in policy and some dramatic rises in economic growth (UNDP 2014). In a developmental sense, the redistributive aspects of growth have been an increasing focus of attention (Maxwell 2003), with rising inequality a critical constraint on poverty alleviation. The question then becomes: what drives inequality, and how does the argument of Picketty and Saez regarding the central importance of institutions and policies translate into Melanesia, with rapidly evolving social and economic contexts produced in the main by resource extraction?

This important collection, based around three chapters derived from the work of the Swiss National Science Foundation–funded project among the Wampar in the Northeast of Papua New Guinea (PNG), provides insights into the micro- and meso-level processes by which individuals and communities in Papua New Guinea seek to capture a share—sometimes what appears as a disproportionate share—of the economic benefits of extractive industries. One of the most frequent observations regarding resource developments in the Pacific—particularly mining—is that they drive the development of 'novel inequalities' among adjacent and affected communities. As Knauft in his chapter (Chapter 6) notes, resource development becomes 'the pinnacle not only of fantasised projection, but also of ultimate inequity bequeathed by modern development'. The chapters here use deep, fine-grained ethnographic material to extend this basic argument in various ways, examining the implications—social, cultural and moral—of these processes and outcomes to better understand the ways in which extraction drives transformation of small-scale societies in PNG. Most of the chapters are focused on the early or pre-stages of extractive projects, where the anticipation of transformations, in a positive sense, are highest and the developing inequalities are at their most novel.

This work, and other recent material on inequality and resource development in Melanesia (see Bainton et al. 2021), speak directly to two common claims regarding large-scale resource exploitation in the region. Firstly, the fact that inequalities arise is a recognition that there are some in the communities who do well from these developments, and that social and economic stratification within the community appears to be a corollary of mineral development. In itself this is an important inflection on broader generalisations about mining (and other extractive industries) as forms of development that bring 'nothing but despair' to indigenous communities in the region (and indeed globally). Clearly there are some in the communities who do well, economically at least, from these developments, whose lot is more than 'despair'. Understanding the processes by which this happens—and how despair can be mitigated or avoided—is clearly a priority for communities and policy makers.

Secondly, most of the chapters focus on the development and influence of leaders within these communities. Big men/brokers/representatives/agents are all forms of leadership—some of which are novel, and some more traditional—that come to the fore around these resource developments. Typically, in both popular and academic work on resource development, labels such as 'elites', 'representatives', 'brokers' or 'middle-men' come with the implication that this stratification and the processes behind it can be easily understood and categorised (as 'elite formation', for example). And there is also usually a dark moral undertone to this discourse, with negative judgements cast on the people involved—'corrupt' is perhaps the most deployed of the adjectives—that implies that these people's position and subsequent wealth comes at the expense of a broader 'grassroots' community. The chapters here illustrate—several directly— the complexity, nuance and dynamic variability that often accompanies these positionalities in the Melanesian context, particularly over time, and especially in the context of resource development.

This afterword picks up on three closely related threads that run through the contributions to the volume, all of which, in settings like PNG, complicate Picketty and Saez's focus on 'institutions and policy': first, issues around land and its paradoxical effects on inequality; second, a further exploration of the issue of representation and leadership as processes that affect inequality within communities; and finally the ways in which forms of communication in Melanesian societies structure the perceptions of inequality in these contexts. I then briefly add a few notes to reinforce these threads from older empirical work on inequality I conducted at the Porgera gold mine in the early 1990s, before revisiting the contribution of this volume to our understanding of inequality in the contemporary world.

Land as a Central (Melanesian) Organising Frame

To pick up Picketty and Saez's 'institutions and policies' in the Melanesian context, explicit expressions of these directed at inequality are (for the most part) absent, largely because like much of the rest of Melanesia, the terrain of inequality is a 'policy-free' zone. What the chapters in this volume show, though, is that there are a variety of malleable and loosely bounded practices at the local level that produce, maintain and drive inequalities within mine-affected communities, and well beyond.

Most significantly, land sits at the centre of the contextual frame within which novel inequalities arise and play out. This is because the customary forms of land tenure in Melanesia that cover between 86 and 97 per cent of the land area (depending how you count it) encompass a wide variety of systems across the hundreds of equally varied customary groups that make up the region's social landscape. Land also sits at the core of what some describe as the 'myth' of Melanesian egalitarianism, with a popular political sentiment being that because of people's links to land, no one is poor, or at least no one goes hungry. The close links between traditional 'institutions and policies' around land and the broader norms and structures of traditional societies are a critical component of contemporary patterns and processes of wealth and poverty in these communities. Overlaying these local societal structures—and indeed often reshaping them profoundly—are the contemporary state-based processes that seek to transform tenure arrangements away from 'custom'. These have been a point of contention from before Independence in PNG. There are a number of formal, state-driven pathways by which this can occur (there is not space here to go into the details and differences between Incorporated Land Groups (ILGs), agency agreements, Special Agricultural Business Leases (SABL) and Clan Land Usage Agreements (CLUA) etc., but see Schwoerer's Chapter 2, this volume, and Weiner and Glaskin 2007), each backed by legislation and particular institutional forms. Each of these regularise and formalise patterns of land 'ownership' that, as the authors of these chapters note, have implications for novel social and economic inclusions and exclusions—and the reshaping of relationships between people and land—within these communities.

Resource development brings wealth, but how that wealth (and the cost) is distributed across a population is also tied to the institutions of state (through legislated processes around royalty and other benefit distribution, compensation), the particular nature of the resource project and its land requirements, and the political manipulation of these institutions and rules (mostly rules about land) by the power dynamics that exist within the local society. It is these structures and interplays of power—intimately connected to land—that are so beautifully explored in the preceding chapters. As Knauft notes, land is the one potentially valuable commodity that many remote Melanesian communities have in the globalised world, hence claims to and ownership of this land are seen as key to creating a 'modern' future. Once this claim-making becomes a trajectory on which some or all in the community embark, all the processes that others in this volume describe start to kick in: the (re)construction of histories, micro-politics, connections and sheer serendipity begin to produce exclusions, inclusions and inequalities.

Tobias Schwoerer (Chapter 2) clearly points to the central importance of land and particularly the opportunities that strategically located land opens up for groups to engage with extraction and, through this, with a (mostly imagined) modernity. The key point he makes, though, is that extractive projects and possibilities become a central means of securing claims to land, through the various legalistic mechanisms and entities he describes: ILGs, CLUAs, VCLRs (Voluntary Customary Land Registration), etc. Loose and evolving groups become fixed entities (Tom Ernst's (1999) 'entification') with their identities tied to discrete tracts of land, in an idealised sense, though really in a practical sense. There is an active generation of 'exclusions' and inequalities through these various legal processes and strategies, typically as a 'by-product' of others pursuing their own agendas.

Leadership, Representation and Brokership

As noted above, one of the key factors that is exposed in the preceding chapters is leadership and its diverse, ambiguous and almost invariably contested nature within the various communities. Who gets to speak, and for whom, is perhaps the central problem of these large-scale resource developments in Melanesia, and the localised examples given in the volume here of 'institutions and policies' around leadership show the variations across Melanesia from one community to another. Indeed, individual aspirations towards leadership or representation reflect a process in which these 'leaders' create an idealised sense of what and how they want 'their' community to be, a process that is constantly evolving and regularly challenged.

In the PNG context, the institutions and policies around these roles sit at the intersection of the 'modern' and the 'traditional'. As noted above, new state forms of corporate groups, each requiring different forms and processes to determine group representation, arise out of the new institutional–legal forms that are brought about by evolving state policy and regulation; hence the ILGs and CLUAs require quite distinct processes and legal forms of representation, and come with differing levels of responsibility to 'represent'. The internal processes in each case—complete with highly contested decisions around who gets to be represented, and by whom—are likely to drive particular patterns and forms of inequality, some of which are likely to change through time, while others will persist. Each of the chapters in the volume has explored aspects of the question of leadership from different perspectives.

Willem Church's chapter (Chapter 3) is implicitly rather than explicitly focused on leadership, as it seeks to develop a model of how inequality becomes embedded in these novel contexts through a process of 'stratifying factional competition'. What this does is link questions of representation and authority back to the question of institutions and policy. The key argument he makes is that the creation and transformation of factions within communities is a competitive process, not between discrete, bounded pre-existing social units present before mining arrives, but rather fractional competition and conflict drives the formation of novel social units, or particular fractional forms. This chapter presents a fine-grained description of legal cases going back over 40 years that illustrates how legal entities and exclusions/inequalities evolved and became increasingly robust and entrenched through engagement in these legal processes. In this context, conflict is as much a driver of the formation of these novel social units as the social units are of conflict: factional competition is then a dynamic feedback process co-creating the very organisations involved in the competition.

The place of the legal systems, the instruments of institutions (ILGs, etc.), and the back-and-forth between traditional and modern forms and positions of leadership then becomes central to the development of forms of inequality. Willem concludes that while corruption and a lack of coordination are often seen as a failure of governance in this space, they can from a local perspective actually represent successful coordination by parties that are benefiting from the status quo.

Meanwhile, Schwoerer (Chapter 2) points to the importance of education for leadership, with typically more educated and often ex-government officers better positioned to more aggressively pursue opportunities for novel forms of wealth. Again, the influence of the state—through education accessibility and quality—plays into leadership and inequality. The inequalities in access and the benefits that these modern leaders are able to secure are not always novel, and often entrench rather than transcend pre-existing enmities and disputes.

Monica Minnegal and Peter Dwyer (Chapter 4) shift the geographic focus, as well as the thematic focus, by providing a beautifully written life history of a leader ('broker' or 'middle-man') that illustrates the processes that both drive and are a consequence of inequality. This account highlights how such liminally positioned brokers/leaders/representatives are at once powerful, morally ambiguous and ultimately vulnerable to alienation, failure and loss. Such positioning means that their actions constantly involve both the navigation and the generation of inequalities. Minnegal and Dwyer also emphasise the role of these leaders/'brokers' in the epistemological and ontological shifts that are transforming Febi and Kubo into 'new kinds of people' that, as they come into being, also generate novel inequalities (and often a profound sense of loss). In a similar focus on the moral dimensions of leadership in such societies, Bettina Beer (Chapter 5) highlights how the claims by leaders of 'giving back' to their communities doesn't deter local evaluatory processes that are highly relational; hence understandings of wealth are often opaque in terms of people being more or less wealthy than others believe.

In sum, then, the leadership contests that emerge, often between long-established (traditional, if you like) forms of leadership and the 'novel' forms of leadership that arise as a result of the structuring and conditioning of communities in line with—or sometimes in opposition to—the legislative, regulatory and judicial needs of the state and resource developers (which are more typically pursued by the better educated and connected members of the community), are critical to shaping the dynamics of community responses and the rise of inequalities.

Communication and Its Manifestations

Finally, in terms of connective themes, the communication of information, and moral messages, is critical to the ways in which locally inflected forms of inequality develop in these contexts. Again, localised 'institutions and policies' dominate in environments where more formal channels and means of communication from central actors are limited, if they are present at all. Formal government communication channels (such as mining wardens and local-level government) are infrequently used and are subject to being co-opted by the demands of local political actors. Hence in Schwoerer's chapter (Chapter 2) is the concern with the ways in which information (and misinformation) flows within communities, between state and corporate actors and communities, to become a 'swirling mass of rumours and expectations, misunderstandings and confusions' creating an information-poor, high-trust decision-making environment. Church's case study (Chapter 3) points also to the ways in which information flows are deliberately restricted and contained by those seeking to capture benefit streams.

Beer's chapter (Chapter 5) is also concerned with information flows and their implications among the Wampar, in the context of evolving inequalities. The unique insight this chapter offers is how inequality reflects and inflects the moral world of Melanesia. It rises as the dark side of development, with a sense of the loss of community and the rupture of the networks of relationships that so completely structure and give meaning to lives in these societies. She highlights how shaming of individuals for the perceived breaking of moral codes around Melanesian sharing can be a means of containing some exclusive forms of wealth accumulation. At other times, much of the evaluatory 'work' in communities is typically based on 'gossip' (and as reflected more broadly in the region, cf. Brison 1992), which can drive conflict, but can also be equally important for developing often ephemeral forms of social cohesion. As Bruce Knauft (Chapter 6) notes, 'actual information and accurate news pales beside the force of cultural and imaginative projection'.

Inequality and Change at Porgera, 1992–2019

In Chapter 6, Bruce Knauft draws on the splendid accounts of change and conflict at the Porgera gold mine in Enga Province by Golub (2014) and Jacka (2015) to illustrate (respectively) the simultaneous social transformations underway: the rising material wealth alongside abject squalor around the mine, and the strong geographic divide between the Ipili in the upper valley (where the mine from which wealth flows is located) and the marginalised but adjacent and socially interconnected lower Porgera areas. These trajectories are long-standing: a household survey and other reports I completed with a team from University of Papua New Guinea at Porgera in the early to mid-1990s (within two years of the mine starting production) already pointed to several axes of 'novel' inequality that were becoming apparent and entrenched. The reports were subsequently published as chapters in Filer (1999). Similar household-level work around the Freeport mine in West Papua in the late 1990s that I was involved in (Banks 2000) mirrored many of the findings discussed below.

This Porgera work identified that:

> In terms of poverty, the total amount of cash flowing into the valley has increased dramatically since the mine became established. On a per capita basis, average incomes are likely to be three to four times higher in real terms within the Porgera area, even given the population growth. But such an assessment conceals massive inequalities within the population at Porgera, inequalities that have increased spectacularly compared to the pre-mine situation. (Banks 2005: 136)

This work identified four overlapping and intersecting axes of inequality that had developed or been heightened by the presence of the mine: geography, hierarchy, gender and residential status (Banks 2005; Jackson and Banks 2002). In terms of geography, the rigidly defined and surveyed mining and infrastructure lease boundaries were critical in driving the development of novel inequalities within the community (more so than the more flexible social boundaries of landowning units). 'Ownership' of strategic pieces of land (the mine site, or the location of associated infrastructure such as camps, waste dumps and even roads) clearly conveyed significant economic and political advantages to some communities. Hence:

> Geographically those groups living within the Special Mining Lease have been at the center of the economic relationship with the company, in particular receiving very large amounts of compensation [for land lost to the mine development] … At a lower level still, within the mining lease, the distribution of compensation money has not been even either, with some clans receiving greater amounts of compensation money as they have lost access to the largest portions of land to the mine development. (Banks 2005:136)

In the Porgeran case, and in contrast to some of the material described in the earlier chapters, the boundaries of the corporate groups ('clans') became fixed at an early stage, but the real contest became one over group membership, exacerbated due to the Ipili form of cognatic descent whereby on paper at least, membership of a group could flow through an individual's mother's or father's line, back at least two generations. This significantly extended the social field of individuals who could make claims to 'belonging' to the affected groups.

Within the residential communities (which, as Golub (2014), Jacka (2015), Burton (2014), myself (1997) and others have described, are far from discrete bounded groups of 'landowners'), there were stark differences in wealth within the population across these geographic domains. Two key factors in this appeared to be gender, and what I described as 'hierarchy'—which referred to leadership and status within these communities (both of which had strong pre-colonial reference points). On the latter, one extreme case was noted:

> for the largest compensation payment made in 1992 (K 520,000), two recognized clan leaders or 'big men' acquired 75 percent of the value of the payment directly, and their children were among others on the distribution list. It is not unusual for big men to appear on the distribution lists of compensation payments for several separate clans. The records of royalty payments to the SML [Special Mining Lease] landowners illustrate the same process. (Banks 2005: 136)

The process evident in Schwoerer's chapter, and indeed throughout this volume, whereby the more vocal and forceful individuals and families are typically able to secure more access to rights and wealth is clearly reflected by similar processes at Porgera. Hence individuals—including some born outside the immediate vicinity—were able to use a combination of negotiating skills, networks, influence and cognatic and affinal rights to claim a share of royalty payments and other revenue streams from across multiple clan territories.

Gender was another critical marker of inequality across the early work. Average female income in the surveys was less than a third of the average male income (Banks 2005). This was resented by most of the women themselves (Bonnell 1999), but is a finding that also matches much of the literature on development in PNG (and indeed historically worldwide). While pre-contact gender-based roles and inequalities have been the subject of much anthropological discussion and may not have been a simple matter, relative inequality in material terms increased for most women around the Porgera mine. Given the more recent reports from Porgera of human rights abuses directed towards women (Human Rights Watch 2011), it is clear that for most women at Porgera, mining has disproportionately impacted on them. Beer's chapter in this volume, with a much more fine-grained analysis, highlights both the varied range of effects of resource developments on the diverse livelihoods of women, as well as the ways in which material changes in circumstances become translated through 'tok' into a much more complex, multifaceted and often contradictory understanding and experience of inequality among these women.

The final axis of inequality explored in the earlier work was the evolving differences between Porgerans and the many more recent migrants who had moved into the area in the post-mining phase. Some of the definitional boundaries between 'local' and 'migrant' were blurred even at that early stage, partly as a result of the movement of people from surrounding areas into Porgera during the extended exploration period (from the 1950s onwards) (Banks 1997; Bainton and Banks 2018). Even using a crude marker of 'recent migrant' ('were they born in Porgera'), though, showed some revealing differences:

> One survey showed that the migrants had an income 50 percent greater than Porgerans over the two week period of the survey (Banks 1999). Breaking down the figures by source of income, the migrants earned a greater amount from mining company wages (K 106 compared to K 44 for Porgerans). In addition the surveys found that as well as generally earning more cash, the migrants had fewer gardens, and were more likely to own a business in Porgera than were the original landowners. (Banks 2005: 137)

Subsequently, a recent (Jinks et al. 2019) revisit of adjoining communities pointed to the continuing presence of inequalities in the community (one household of the 21 surveyed accounted for 28.5 per cent of the fortnightly income in the community adjacent to the SML), and mirrored the findings in relation to migrants and 'locals': 'three of the four households with a fortnightly income of over K2,000 were non-landowners …, while six of the seven with incomes of K500 or under were landowner households' (2019: 28). The ability of these migrant non-landowner households to generate incomes greater than locals has long been a point of contention and conflict. It is one of the key reasons too why people move to be closer to these resource projects. As Knauft (Chapter 6) notes: 'most people in rural PNG fantasise and project, but more practically and immediately they yearn … for the amazing wealth they perceive in other areas beyond their own'. And it is a far-reaching effect, as a 'sense of being left behind or left out is often at the heart of conflicts and disputes

… in areas far distant' (Knauft this volume) from the resource extraction sites; indeed they can be sparked, as Knauft notes, by connections to and movements towards distant *possible* extraction projects.

As a result of the patterns of geographic, social inequality noted above, the early survey work found substantial income inequalities within the community. Hence 10 per cent of households in an early survey earned 59 per cent of the total sample income reported, while the lower 50 per cent of the sample earned just 2 per cent of the total income (Banks 1999): 'Eighteen percent had received no money over the fortnightly period, and another 10 percent less than K 10' (Banks 2005: 137). The cash income then was stratifying the community into novel configurations of households, individuals and larger groups (the putative Porgeran 'sub-clans').

The analysis also argued that there were two other factors at work that influenced both the extent of material inequality and its evaluation by people within communities (reflecting the difference that Knauft notes between inequality and inequity). First:

> despite the broad geographic patterns noted, it was apparent from the surveys that there are more marked differences within each of the Porgeran communities than between them, reflecting the influence of hierarchy, gender, and residential status. (Banks 2005: 137)

Second, and following from this, the early survey work also pointed to another key influence within these communities, arguing that while '"trickle down" does not necessarily provide more egalitarian outcomes among traditional societies than in economist's models of modern ones' (Banks 2005: 137), there was evidence that redistribution was occurring to an extent within the affected populations. Colin Filer (1990) had famously commented in relation to the Bougainville crisis that a lack of distribution of mine-derived revenue streams within Melanesian societies was due to the fact that they simply had no appropriate traditions to draw on when it came to distributing this novel wealth. Traditional mechanisms for the distribution of the objects of customary trade and exchange (pigs, pearl shells, dog teeth, etc.), could not, according to Filer, be expected to provide a basis for the equitable distribution of millions of kina in cash (examples of which have been noted above). The sheer scale of this novel wealth then, produces tensions that traditional ways of balancing and managing relationships simply cannot handle. The Porgera surveys (both the historical and the more recent one) showed that despite Filer's caution here, there are still traditional norms of distribution that operate to mitigate some of the material inequalities that mine wealth generates. Hence:

> The most common source of income for individuals in the surveys was from household members or other wantoks and indeed for a quarter of the sample this was the only source of income (Banks 1999). Although the sums involved were often small, there was an element of redistribution occurring within the community. (Banks 2005: 138)

Likewise there were elements of redistribution evident within the exaggerated forms of traditional exchange that appeared at Porgera—bride price and compensation payments increased exponentially, and mining revenues underpinned this growth. Bainton (2010) found the same with mortuary feasting events on Lihir. Societal values of obligations to care for kin still functioned here, although they typically did little to smooth out the extremes in inequality that had been created, and often nurtured culturally specific forms of dependency. Dependency in the modern Western sense carries a heavy moral load, with the neoliberal ethos of individualistic pursuit of wealth and happiness demonising dependency and forms of interconnected communal constructions of society. In Melanesian societies where relationality is central, the relationship between dependency and reciprocity is more complex. Indeed, as is made obvious in Beer's, Knauft's and Minnegal and Dwyer's chapters, the moral load is often on those with wealth to accept and maintain relationships of dependency with kin and others. Much of the clamour and social chaos around resource projects in Melanesia results from the active reshaping of relationships and the values tied to them, and the calculated drive to establish relations of dependency—as a foothold, or a more permanent state (cf. Ferguson 2015 in the African context)—is a significant component of this. Stress and fracturing of norms (Melanesian societal institutions, if you like) of kinship, reciprocity and dependency (see Beer's chapter) in these contexts can shatter the moral codes that tie these communities together.

As the Porgera case above shows, relative inequalities are critical to development prospects, even where absolute wealth increases and absolute poverty drops within communities. The socially determined and geographic inequalities collide with, build on and articulate with polarising processes within these various communities to construct a complex web of material and discursive and different segments of the community, and internally between different inequities. In this context, as Knauft notes in Chapter 6, the 'abrogation of meaningful reciprocity' (between the developers and differently positioned individuals within the community) sit at the heart of the processes of 'social disintegration' (Filer 1990), tensions and conflicts.

There is a flip side to the wealth that does flow unequally into these communities, and the inequalities that result. The agreements and policies that sit behind the various revenue streams direct the bulk of the compensation and royalty flows to individuals and communities that have had their land altered, destroyed or alienated by resource development, alongside other forms of transformation and dislocation in their lives. In much of the discourse around wealthy, greedy, rapacious resource landowners, there is often insufficient recognition by observers (company, government, the broader public and many academics) that the mining leaseholders' wealth has come at a cost. To talk (as economists often do) of royalties and dividends as 'unearned resource rents' is to belittle the profound cultural and emotional loss that typically accompanies the loss of land and, as explored in the chapters above, the shifts within society that accompany mining development.

Conclusion: Inequality, Institutions, Policy and Nuance

As is obvious from the chapters in this volume, the disparities generated through the unequal distribution of revenues, and the interplay with group formation and local identity politics is a key driver of conflict around resource extraction projects. The chapters here productively shift the focus of the debate around inequality from an individualistic/societal one (the frame within which Picketty and Saez's opening statement arises) to one constructed around a more Melanesian form of relationality. Not only is 'relative inequality' (or inequity, as Knauft notes) a key concept, but more fundamentally a relational view of the world, based on an understanding that foregrounds relationships—between people, land, materiality and the spirit world—forces us to rethink the processes and levers that frame and construct inequality. As Beer notes:

different segments of local social fields can or cannot engage the encompassing global processes to different degrees and in different ways, depending on a host of social [and economic] factors. The upshot of such initial differential processes is frequently the production of significant social and economic demarcations, which itself is crucial to the generation of more entrenched social contracts in the medium and longer term. (Beer, this volume)

The ethnographic focus on practices is able to be used so productively within this volume to interrogate broader discussions of inequality precisely because these relationships across multiple levels and contexts are so fine-grained, complex and diverse. Any worthwhile examination of inequality within the evolving global systems needs to pay heed to the nuance of cultural contexts, relationships and lived practices, not just national institutions and policy.

References

Bainton, N., 2010. *The Lihir Destiny: Cultural Responses to Mining in Melanesia*. Canberra: ANU E Press (Asia-Pacific Environment Monographs). doi.org/10.22459/LD.10.2010

Bainton, N. and G. Banks, 2018. 'Land and Access: A Framework for Analysing Mining, Migration and Development in Melanesia.' *Sustainable Development* 26(5): 450–460. doi.org/10.1002/sd.1890

Bainton, N. and D. McDougall, 2021. 'Unequal Lives in the Western Pacific.' In N.A. Bainton, D. McDougall, K. Alexeyeff and J. Cox (eds), *Unequal Lives: Gender, Race and Class in the Western Pacific*. Canberra: ANU Press. doi.org/10.22459/UE.2020.01

Bainton, N.A., D. McDougall, K. Alexeyeff and J. Cox (eds), 2021. *Unequal Lives: Gender, Race and Class in the Western Pacific*. Canberra: ANU Press. doi.org/10.22459/UE.2020

Banks, G., 1997. Mountain of Desire: Mining Company and Local Community at the Porgera Goldmine, Papua New Guinea. Canberra: The Australian National University (PhD thesis).

———, 1999. 'The Economic Impact of the Mine.' In C. Filer (ed.), *Dilemmas of Development: The Social and Economic Impact of the Porgera Gold Mine, 1989–1994*. Canberra: Asia Pacific Press.

———, 2000. 'Social Impact Assessment Monitoring and Household Surveys'. In L. Goldman (ed.), *Social Impact Analysis: An Applied Anthropology Manual.* Oxford: Berg.

———, 2005. 'Globalization, Poverty, and Hyperdevelopment in Papua New Guinea's Mining Sector.' *Focaal* 46: 128–144. doi.org/10.3167/0920129067 80786799

Bonnell, S., 1999. 'Social Change in the Porgera Valley.' In C. Filer (ed.), *Dilemmas of Development: The Social and Economic Impact of the Porgera Gold Mine, 1989–1994.* Canberra: Asia Pacific Press.

Brison, K., 1992. *Just Talk: Gossip, Meetings, and Power in a Papua New Guinea Village.* Berkeley: University of California Press. doi.org/10.1525/california/ 9780520077003.001.0001

Burton, J., 2014. 'Agency and the "Avatar" Narrative at the Porgera Gold Mine, Papua New Guinea.' *Journal de la Société des Océanistes* 138–139: 37–52. doi.org/10.4000/jso.7118

Ernst, T., 1999. 'Land, Stories, and Resources: Discourse and Entification in Onabasulu Modernity.' *American Anthropologist* 101(1): 88–97. doi.org/ 10.1525/aa.1999.101.1.88

Ferguson, J., 2015. *Give a Man a Fish: Reflections on the New Politics of Distribution.* Durham: Duke University Press. doi.org/10.2307/j.ctv1198xwr

Filer, C., 1990. 'The Bougainville Rebellion, the Mining Industry and the Process of Social Disintegration in Papua New Guinea.' *Canberra Anthropology* 13(1): 1–39. doi.org/10.1080/03149099009508487

——— (ed.), 1999. *Dilemmas of Development: The Social and Economic Impact of the Porgera Gold Mine, 1989–1994.* Canberra: Asia Pacific Press.

Golub, A., 2014. *Leviathans at the Gold Mine: Creating Indigenous and Corporate Actors in Papua New Guinea.* Durham: Duke University Press. doi.org/10.1515/ 9780822377399

Human Rights Watch, 2011. 'Gold's Costly Dividend: Human Rights Impacts of Papua New Guinea's Porgera Gold Mine.' Human Rights Watch. Viewed 22 July 2021 at: www.hrw.org/report/2011/02/01/golds-costly-dividend/ human-rights-impacts-papua-new-guineas-porgera-gold-mine

Jacka, J.K., 2015. *Alchemy in the Rain Forest: Politics, Ecology, and Resilience in a New Guinea Mining Area.* Durham: Duke University Press. doi.org/10.1215/ 9780822375012

Jackson, R. and G. Banks, 2002. *In Search of the Serpent's Skin: The History of the Porgera Mine*. Port Moresby: Placer Niugini.

Jinks, B., R.M. Bourke and G. Banks, 2019. *Agriculture, Income, Expenditure and Business in Panandaka and Pakien*. Canberra: ANU Enterprise, for the Porgera Joint Venture.

Maxwell, S., 2003. 'Heaven or Hubris: Reflections on the New "New Poverty Agenda".' *Development Policy Review* 21(1): 5–25. doi.org/10.1111/1467-7679.00196

Picketty, T. and E. Saez, 2014. 'Inequality in the Long Run.' *Science* 344 (6186): 838–843. doi.org/10.1126/science.1251936

UNDP (United Nations Development Programme), 2014. *National Human Development Report Papua New Guinea: From Wealth to Wellbeing: Translating Revenue into Sustainable Human Development*. Port Moresby: UNDP.

Weiner, J. and K. Glaskin (eds), 2007. *Customary Land Tenure and Registration in Australia and Papua New Guinea: Anthropological Perspectives*. Canberra: ANU E Press (Asia-Pacific Environment Monographs). doi.org/10.22459/CLTRAPNG.06.2007

www.ingramcontent.com/pod-product-compliance
Lightning Source LLC
Chambersburg PA
CBHW050809270326
41926CB00026B/4644

* 9 7 8 1 7 6 0 4 6 5 1 8 6 *